# WEATHER

## IN THE

# COURTROOM

MEMOIRS FROM A CAREER IN FORENSIC METEOROLOGY

William H. Haggard

AMERICAN METEOROLOGICAL SOCIETY

Front cover photograph: Fuse/Corbis/Thinkstock

Published by the American Meteorological Society
45 Beacon Street, Boston, Massachusetts 02108

The mission of the American Meteorological Society is to advance the atmospheric and related sciences, technologies, applications, and services for the benefit of society. Founded in 1919, the AMS has a membership of more than 13,000 and represents the premier scientific and professional society serving the atmospheric and related sciences. Additional information regarding society activities and membership can be found at www.ametsoc.org.

Library of Congress Cataloging-in-Publication Data

Names: Haggard, William H., author.
Title: Weather in the courtroom : memoirs from a career in forensic
  meteorology / William H. Haggard.
Description: First. | Boston, Massachusetts : American Meteorological
  Society, [2016]
Identifiers: LCCN 2016032887 | ISBN 9781940033952 (pbk.)
Subjects: LCSH: Crime and weather. | Forensic sciences.
Classification: LCC HV6030 .H34 2016 | DDC 363.25 [B] —dc23
LC record available at https://lccn.loc.gov/2016032887

FSC
www.fsc.org
MIX
Paper from
responsible sources
FSC® C005010

Dedicated to the memory of Martina

# CONTENTS

# INTRODUCTION

Weather has always been my passion. With a pending degree in meteorology from the Massachusetts Institute of Technology (MIT) in 1942, I joined
the U.S. Navy and was assigned to weather-related duties ashore and at sea
during the Second World War. After the war, I obtained my master's degree
in meteorology from the University of Chicago and became involved in a
series of forecasting jobs in the U.S. Weather Bureau. The navy recalled me
for three years during the Korean conflict after which I worked in the Office of Climatology. These were exciting years of scientific discovery. I was
privileged to work with meteorological leaders, who encouraged my passion
for weather. The American Meteorological Society (AMS) held conferences
and offered publications that kept us abreast of the rapid developments of
our field. In 1961, I applied for the position of deputy director of the National
Climatic Data Center (NCDC) in Asheville, North Carolina, and was hired.
The center archives climatological data from all over the world and makes
it available to the public. While serving as the director of NCDC between
1963 and 1975, I was impressed by the large number of attorneys requesting
weather data for their litigation cases. The center would offer blue ribbon
and gold sealed data certified by the Department of Commerce that could
be submitted as evidence in a court of law, but government meteorologists

could not be released from their full time duties to interpret this data in the courtroom. This void was filled by consulting meteorologists in the private sector.

The AMS had, in the meantime, developed a certification program to authenticate the qualifications of consulting meteorologists. Certified Consulting Meteorologists (CCMs) are tested and certified by the Society to excel in their profession with high ethical standards and technical competence. I took the challenging test and was the 150th to obtain my CCM. I had already decided that after retirement from the federal government I would become a consulting meteorologist specializing in forensic work. This specialty would allow me to bring weather testimony into the courtroom. Hired by attorneys as an expert witness in forensic meteorology, I would study the available past weather data pertaining to the date and location of the event and then, using my education, years of job experience, and knowledge, form an opinion based on the facts and present it to the court.

In 1976, I formed the Climatological Consulting Corporation (CCC) and sent letters of introduction to lawyers listed in legal directories who appeared likely to need weather testimony in their litigation. They included aviation, marine, and insurance lawyers. The first response came several months later from Charles Hagans in Anchorage, Alaska, who needed an expert weather witness in a plane crash. Though the weather analysis was complex, I was faced with the great challenge of learning courtroom procedures and the behavioral techniques that would convince the jury that I was a credible witness. I learned the importance of effective visuals with the concept that people retain 80% of visual information versus 20% audible. This basic education was a great foundation to become a better expert witness. I later studied books such as *The Expert Witness Handbook* by Dan Poynter (Poynter 1987), *The Scientist and Engineer in Court* by Michael Bradley (Bradley 1983), and *The Role of the Expert Witness in a Court Trial* by Benjamin Cantor (Cantor 1996).

After a slow start, word of mouth advertising, attending various legal conferences, and hard work all helped increase my business. Meeting and working with so many talented attorneys and staff was a great experience. My staff increased, my travel increased, and my lifelong passion of working with the weather continued. We worked on hundreds of cases varying from simple "slip and falls" to complex weather patterns such as those found in the Perfect Storm of 1991. As the caseload increased, I realized the need for more specialized interpretation of satellite data, hydrology, radar, and severe weather. I began affiliating with other CCMs in these subspecialties to better meet the client's needs. These associates greatly assisted me in their specialty and enhanced the final work product.

The purpose of this book is to share some of these significant cases in which I played a part. With so many from which to select, one goal was to choose interesting cases, and another was to show the variety of weather situations that can affect litigation. Each chapter recounts a different weather phenomenon: snowstorm, microburst, cloud cover, ocean storm, rain squall, flood, drought, hurricane, fog, tornado, rain/hail storm, temperature, or wind. The lawsuits resulted from their impact on aviation, marine crafts, highways, industrial structures, buildings, recreational parks, and land management. The book opens with my first case.

# CHAPTER ONE

# BROWN V. JONZ

*The Mysterious Loss of Two Congressmen, Alaska, 1972*

## THE EVENT

Two congressmen, an administrative assistant, and pilot disappeared while flying from Anchorage to Juneau, Alaska, on October 16, 1972. Alaskan Congressman Nick Begich was campaigning for reelection to the 93rd Congress. House Majority Leader Hale Boggs, from Louisiana, was in Alaska to support Begich in his reelection campaign. Both had spoken at a fundraising meeting in Anchorage on October 15 and were to attend a campaign rally in the capital, Juneau, the next day.

Don Jonz, a supporter of Congressman Begich's campaign, was the chief pilot and sole stockholder of Pan Alaska Airways, Ltd., an Alaskan air taxi service. He offered to fly the three men to Juneau for free in his company's Cessna 310C, as a convenience to them and a political favor. His flight experience included many hours of cross-country, multiengine, and instrument flight time (flight time with reliance on the instruments within the aircraft rather than visual reference to the environment). He had flown the Cessna 310C, a twin-engine, propeller-driven, multipassenger aircraft, from its base in Fairbanks to Anchorage on the evening of October 15 and remained overnight in Anchorage in preparation for the next day's 645-mile flight to Juneau.

## PLANNING THE FLIGHT

At 6:56 a.m. Alaska daylight time on October 16, Pilot Jonz telephoned the U.S. Government Flight Service Station (FSS) at Anchorage, which provides weather- and flight-related information to pilots, and asked for the existing and forecast weather along the route of flight from Anchorage to Juneau for the next six hours, including Anchorage, Cordova, Yakutat, Sitka, and Juneau weather details (see Figure 1.1). The weather briefer provided the 6:00 a.m. weather details at those locations as well as the forecasts for those airports: the forecast winds from the surface to 12,000 feet, the area forecasts (a prediction of weather along the entire route), as well as a prediction of the conditions within a major mountain pass, Portage Pass, 36 miles southeast of Anchorage, through which the plane would have to fly. The weather briefing indicated that at 6:00 a.m. there were portions of the route that would require instrument flight rules because of early morning clouds and fog (see Figure 1.2).

## THE FLIGHT

After the flight departed at 9:00 a.m., the pilot called the FSS to report his planned route of flight and estimated time of flight of 3 hours and 30 minutes. He received updated weather information, indicating improving conditions during the later morning hours in the Yakutat to Juneau portion of the flight when the fog and low clouds would be burned off by the warming sunlight. The major mountain pass at Portage was forecast to be closed for visual flight rules (VFR flight) by the National Weather Service (NWS) at Anchorage, based on an estimate of the likely low clouds in the valley, though there were no observing stations in the valley. Pilots can file a VFR plan under good weather conditions, where they can see where they are going. Otherwise they are required to file and fly under instrument flight rules (IFR) under bad weather conditions, where clouds, fog, and precipitation can obstruct their sight, requiring instruments for their orientation.

The pilot radioed the FSS and filed a VFR flight plan from Anchorage to Yakutat then direct to Juneau, his destination. The specialist accepted this plan and asked if he had emergency gear and a locator beacon [electronic locator transmitter (ELT), which sends radio signals activated in crashed aircraft] aboard. He replied "affirmative." The plane was last seen by Anchorage Airport Tower personnel about two miles southeast of the airport at 2,000 feet, headed toward Portage Pass. Four hours and 15 minutes later, the U.S. Coast Guard Rescue Coordination Center (RCC) at Juneau advised the U.S.

Air Force RCC at Elmendorf Air Force Base (north of Anchorage) that the Cessna flight was 45 minutes overdue at Juneau.

## THE SEARCH

The Air Force base contacted all airfields near the route for any information they might have about the flight. None was found. They then diverted an already airborne, Lockheed HC-130, four-engine, turbo-propeller, specially equipped search and rescue aircraft to search along the planned route of flight for the missing aircraft. It failed to find any evidence of a downed plane. A massive search followed from October 16 to November 24, covering more than 300,000 square miles. Aircraft and helicopters flew more than 3,000 hours. I was told that the Lockheed SR71 Blackbird, a secret military plane, was included in the search but without results. Extensive marine searches by ship of Prince William Sound, the Gulf of Alaska, and Icy Strait were conducted. Ground personnel twice searched much of Portage Pass, which is where some believed the plane had likely crashed. Despite the thoroughness of these extensive searches, nothing was found that could be identified as related to the plane or its occupants.

## THE NATIONAL TRANSPORTATION SAFETY BOARD

The National Transportation Safety Board (NTSB) conducted an investigation of the accident and issued an "Aircraft Accident Report" (NTSB-AAR-73-01) (NTSB 1973), which was adopted on January 31, 1973. The 25-page report, an official government document available to the public and widely distributed, followed a standard format of abstract, investigation, analysis, conclusions, probable cause, recommendations, and had five attachments dealing with proposed route of flight, aviation weather forecasts, pilot information, aircraft information, and investigation and hearing. In describing the 38-day search, the NTSB made no mention of the role of the U.S. Air Force Blackbird, probably because of the secret classification of the plane at, or prior, to that time. The NTSB report's four conclusions were as follows:

- The pilot was qualified.
- The aircraft was qualified, though it was not equipped with an emergency locator transmitter.
- The pilot did not have a portable emergency locator transmitter nor survival equipment on board.

- The weather conditions along the route were inappropriate for visual flight rules.

Under the section of the report titled probable cause, the NTSB report stated: "The Safety Board is unable to determine the probable cause of this accident from the evidence presently available. If the aircraft is found, the Safety Board will continue the investigation and make a determination as to the probable cause of the accident" (NTSB 1973, p. 9).

## ODD REPORTS

Not mentioned by the National Transportation Safety Board in its factual report, I was told that news articles contained a number of odd occurrences stating that a Coast Guard helicopter pilot reported picking up very strong emergency locator beacon signals the day after the flight. Pilots of two jet-powered Coast Guard helicopters said they picked up a signal 10 miles west of Juneau in mountainous terrain. With no other aircraft reported missing at that time, who else might have transmitted such signals? Another odd report came from a story of five California citizen band radio operators, who were convinced they had talked to the missing pilot. The information they relayed to the Air Force in their interview was ignored.

## THE SUIT IN THE SUPERIOR COURT OF ALASKA

The families of the two congressmen had accepted the $150,000 per seat insurance carried by Pan Alaska Airways and agreed not to file lawsuits. However, the widow of the administrative assistant Russel Brown, also on board the flight, sought greater compensation, declined the insurance, and filed a suit against the pilot's estate. She alleged her husband's wrongful death resulted from negligence and improper actions of the pilot Don Jonz. Her case hinged on the legality of his flight as filed. The plaintiff wanted to prove the weather was too bad for pilot Don Jonz to take off and fly under a visual flight rule plan. Visibility in Portage Pass was the main focus of the suit.

Charles Hagans was the attorney defending the estate of pilot Don Jonz. He realized it was a weather-related case. In February 1976 he sought my aid as an expert witness, professionally qualified in aviation weather, to assist in his defense of the estate of the deceased pilot. He asked that I obtain the pertinent weather data from the National Climatic Data Center and analyze the weather conditions along the planned route of flight throughout the day of the flight. He requested I form an opinion as to whether VFR flight was possible. My analysis relied on observed and forecast weather conditions and what information would have been available to the flight service station briefer in his conversation with the pilot.

## PORTAGE PASS

After the trial started, Charles Hagans chartered a small aircraft and pilot. He wanted me to be flown over the early portion of the flight route to become familiar with the terrain and the weather. We flew over Turnagain Arm of Cook Inlet (see Figure 1.3). At the east end of the water-filled arm is Portage Pass, a valley bottom at 400 feet above sea level running northwest to southeast between the 3,000- and 6,000-foot high mountains of the Chugach Range, through which small vessels from Prince William Sound had been portaged to reach Anchorage in the early history of settlement of the area.

The valley is considered open to VFR flight when the skies within the pass are clear and visibility good. If clouds cover more than half the sky below 8,000 feet, the valley is considered open only if the bottoms of those clouds are more than 3,000 feet above the ground and both sides of the valley and its pass are clearly visible from an aircraft below the clouds. Otherwise the valley is considered closed. Without observing stations, pilots were permitted to take a look to determine the conditions within the pass and proceed if they were good, but return from the valley before entering clouds.

On both the day of the Cessna 310C flight in 1972 and my flight in 1976, the prediction was that Portage Pass was closed, though after taking a look on my flight in 1976, it was open with only a few scattered clouds above 3,000 feet above the ground. This flight became an excellent and spectacular sightseeing trip during which the pilot pointed out multiple remnants of crashed planes on the lower slopes of the pass and recited the names of their pilots and when the crashes had occurred. None, of course, were the Cessna 310C. The familiarization flight continued southeast over Prince William Sound and returned to Anchorage without incident.

## COURTROOM LESSONS

I quickly learned that to be effective, in addition to his/her expertise in their professional field of work, an expert witness needs to be aware of the fundamentals of courtroom procedure. Since I was, at that time, inexperienced in courtroom procedures, Attorney Hagans had me sit quietly in the spectator area of the courtroom during the plaintiff proceedings, observing and noting the weather testimony of the several Alaskan bush pilots. They all testified Don Jonz was negligent in attempting to fly under VFR in the existing weather and should have never attempted the flight even though he was IFR rated.

The plaintiff's case took several days, during which I worked in the evenings with Sanford Gibbs (then a clerk and later a partner in the law firm). His probing questions and enthusiastic interest in weather provided a challenging opportunity for me to obtain greater knowledge of the complexities of Alaskan weather. He spent many hours with me going over details of the anticipated weather testimony and developing proper direct examination questions. The direct questions, which are standard courtroom procedure, are those asked to the witness by the engaging attorney to illustrate the facts. These are followed by the cross-examination questions, where the opposing attorney attempts to dispute the testimony.

We also coordinated the improvement of courtroom visuals to illustrate the anticipated weather testimony. Though I had prepared some overly simplistic visuals at my office, I was encouraged to engage a graphic art firm to prepare illustrative visuals to demonstrate the relation of the observed and forecast weather pertinent to the flight and its relation to the complex Alaskan topography along the planned route of flight. The personal guidance by Sandy Gibbs and Charles Hagans was extremely valuable to me in preparation for my first time on the witness stand. Such advice was as follows:

- Tell only the truth.
- Listen to the questions, and answer only the questions asked.
- Do not volunteer information not directly related to the question asked.
- If a yes or a no will suffice, simply say yes or no.
- Never answer with a nod or a head shake; always say yes or no.
- Wait for the entire question before answering.
- Think carefully before speaking.
- When answering, look directly at the judge, or the jury, if present (do not look at the questioning attorney).
- NEVER lose your temper or argue with the questioner, and remain calm and polite at all times.
- If an attorney says "objection," stop answering, wait for the judge to rule, and answer only the new question.
- If you do not know, say so (do not speculate).
- Be as brief as possible in each answer.

- Speak loudly enough that all the participants (judge, jury, attorneys, recorder, etc.) can hear you.
- Be polite at all times, even if you feel an attorney is badgering you.
- Never speak to a juror outside the courtroom during the trial.

The hardest for me was the admonition to look at the judge (with a nod) and then speak to and look directly at the jury when answering a question, as my instinct was to look and speak directly to the questioner. Charles Hagans had to move to the side of the jury box, point to them, and say, "Please tell the jury."

In preparation for the defense weather testimony, Sandy Gibbs and I reviewed the National Transportation Safety Board Accident Report; the factual report (everything but the "probable cause") was admissible in litigation and was heavily relied upon by all parties in the case of *Brown v. Jonz*. Since news articles are considered hearsay, they are not admissible as evidence in court. The NTSB Accident Report did not mention the odd reports, so they were not taken into consideration by the court.

As is customary, the plaintiff case was presented in the first several days of the trial. A series of fact witnesses were presented and related the pertinent known facts of the events, including most of the information in the factual portion of the NTSB Accident Report as well as the details of Russel Brown's life and finances and criticism of Don Jonz. These were followed by a number of Alaskan pilots who described their interpretation of the weather on the date of the accident, fight rules, their experiences in flying all or portions of the flight route at various times and under various weather conditions, and why they thought the flight should never have been attempted. Each was asked if they would have attempted to fly the route on a VFR flight plan on the day of the accident. All replied no for reasons such as

- the ceiling and visibility at Juneau at 6:00 a.m. (when Don Jonz obtained his weather briefing) were too low to be able to land there (at that time of day),
- Portage Pass was forecast to be closed, or
- it was a bad weather day.

None compared the weather at specific locations and times when the flight would have passed them with the flight rule requirements, and the plaintiff did not produce a meteorologist to testify but relied upon the several pilots to imply the illegality of the flight under visual flight rules.

When it was time for the defense's weather testimony, Mr. Hagans first had me present my educational and employment history and qualifications and asked the court to accept me as an expert in aviation meteorology. Then he began the direct examination. I was asked to answer a series of factual questions regarding the observed and probable weather along sequential sections of the route. I based my answers on official certified copies of data obtained from the government weather archives at the National Climatic Data Center.

The newly created visuals, including a map of the geography of Alaska between Anchorage and Juneau, with the locations of the weather reporting stations, airways, flight route, mountain passes, and so on, were placed in the courtroom to permit the judge and jury to view them continuously during the defense weather testimony. The whole purpose of my testimony was to show the route segment by segment over the timeline of the flight, demonstrating the progressive VFR conditions as the flight would have progressed. This was followed by a number of questions regarding my own observations on the familiarization flight through Portage Pass and over Prince William Sound. The cross-examination was lengthy, with questions designed to cast doubt on the pilot's ability to maintain visual ground reference during all phases of the flight.

## THE VERDICT

At the conclusion of the questioning of all witnesses, the jury retired and considered their verdict. Under the rules of the court, at no time was the jury informed of the existence or availability of the insurance carried by Pan Alaska Airways or of Mrs. Brown's having declined such payment. They returned a verdict in which Mrs. Brown was awarded less than half the amount of the insurance she had declined. They evidently compromised and believed Don Jonz was somehow negligent but not so seriously so as to warrant a large financial award. The plaintiffs appealed to the Alaska Supreme Court, which denied the appeal.

## LESSONS LEARNED

We will never know what really happened. The rate of aviation accidents over vast expanses of rugged terrain continued to challenge the FAA long after the 1972 crash of Don Jonz's Cessna 310C with its congressional passengers. An "aftermath" article in the August 2009 issue of *Flying Magazine* (Garrison 2009) included the statistic that the National Transportation Safety Board has investigated 33 Alaskan aircraft accidents involving 96 fatalities in the years of 2000 through 2008, many involving aircraft that were never found. To counteract these kinds of statistics, a major effort to provide valuable flight safety information began early in the twenty-first century with a decision to install cameras viewing frequently traveled critical locations. Photo images are placed on the internet (Federal Aviation Administration 2015) and can be retrieved online, permitting pilots to see existing conditions before embarking on a flight. By December 2007, 82 operational sites had been installed when the FAA decided to invest $102 million more over the next 26 years for the additional installation and continued operation of aviation cameras in Alaska; 139 more aviation camera sites increased the network to a total of more than 220 sites by the year 2014 (Federal Aviation Administration 2015). At Portage Pass, two cameras were installed near the Prince William Sound at the end of the pass, with one facing southeast while the other faces northwest.

At every site, camera images are generally updated every 10 minutes. When accessed, they may be compared with a clear day image, which contains notation of distances from the camera to prominent geographic features, such as shorelines, mountain peaks, points of interest, and so on. These website images are designed as an FAA supplemental product for situational awareness only. They are, however, of great value in assessing the current conditions at critical locations. The program is designed to prevent unnecessary fuel consumption, reduce aviation accidents, and save lives.

## PERSONAL BENEFITS

This was a great initiation into the courtroom and into my new life. Charles Hagans was satisfied with the work I did and recommended me to other lawyers. I got other cases, including a huge case in which I worked for the government over the legality of building a cross-wind runway at the Anchorage Alaska Airport, which required three years of complex work and testifying in the federal court in Anchorage on behalf of the government. Charles Hagans also encouraged me to attend the Southern Methodist University Air Law Symposium in Dallas, Texas, where more than 700 lawyers met annually, with educational lectures and case histories where I met future clients.

# CHAPTER TWO

# NEVIN V. USA

*The Secret Test That Killed, San Francisco, California, 1950*

## THE EVENT

In early November 1950, 75-year-old Edward Nevin, a retired Pacific Gas and Electric Company worker, died unexpectedly in the Stanford University Hospital in San Francisco, California. He had been recovering so well from a prostate operation he had been sent home, not far from his doctors, to convalesce. His condition suddenly worsened so rapidly he was readmitted to the hospital, where he died. The doctors at the hospital were baffled by his unexpected death. They were further surprised by the sudden and simultaneous illness of 10 other patients, who had undergone recent urological surgeries and became very ill with a serious unidentified bacterial infection.

The doctors, who insisted on an autopsy of Edward Nevin, found his death was caused by an infection of *Serratia marcescens* bacteria, which had attacked his heart valves. *Serratia marcescens* was previously not known to be pathogenic. The doctors were further amazed to find that, when tested, the other 10 suddenly ill patients were all suffering from *Serratia marcescens* infections. This mysterious outbreak was so unusual that a medical group wrote it up for an article in the American Medical Association's *Archives of Internal Medicine* titled "Infection Due to Chromobacteria" (Wheat et al. 1951). There were no further cases and the sudden outbreak remained a mystery for 20 years.

In 1970, government documents about U.S. Army biological warfare tests were declassified and revealed that in September of 1950, the army had conducted a major series of germ warfare tests in the United States. The details of the tests conducted at San Francisco were documented in Special Report 142 (U.S. Chemical Corps Laboratories 1951).

## PUBLIC AWARENESS

In 1976, Special Report 142 was discovered by a reporter for the *San Francisco Chronicle* (Cummings 1976). He published a story tying the death of Edward Nevin and the San Francisco military biological warfare tests together. This newspaper article alerted the Nevin family of the possible cause of the untimely death of their ancestor, and the family members gathered together to discuss the startling discovery. Edward Nevin's grandson, Ed Nevin III, was a medical malpractice attorney in San Francisco; the other family members asked him if the army could be found guilty of their ancestor's death.

Very divided in their opinions over what to do, they finally agreed that Ed Nevin, who was planning a trip to Washington, D.C., should see what he could learn about the army tests and the illnesses and death. Ed Nevin visited congress, senate, and legislative staff members but learned very little. When he met again with the family members in San Francisco, he reported that it would require a lawsuit to learn about the tests. They finally agreed both they and the public had the right to know what had occurred and that a suit for a huge newsworthy amount should be sought. They agreed on $11 million, to make the suit newsworthy.

Ed Nevin filed suit in federal court in 1977. The U.S. attorney in San Francisco immediately requested the claim be summarily dismissed. Federal Judge Samuel Conti denied the request for dismissal, and the case headed for court and trial, with a trial date set for 1979, which would be postponed several times. In preparation for the trial, Ed Nevin requested all documents related to the San Francisco tests and received Special Report 142, which described them in detail.

## THE TESTS

A series of six tests were conducted in 1950. In each of these, bacteria (believed to be harmless) were sprayed from a military ship two miles upwind and offshore, west of the city along a north to south track. Receptors to detect the bacteria were placed throughout the city to determine the pattern of their dispersion over the city. Ed Nevin realized he would need the testimony of a weather expert to analyze the wind patterns during each test. He contacted the American Meteorological Society, in Boston, Massachusetts, for a list of board-certified consulting meteorologists and the American Bar Association seeking information on meteorologists who had testified in weather-related litigation and found my name on both lists. He contacted me in late 1979 and asked that I review Special Report 142 and obtain the required weather data. I was to reconstruct the pattern and density of the clouds of bacteria as they traveled over the city in each of the six tests.

## THE WEATHER ANALYSIS

The six tests were conducted on late afternoons in September of 1950, in which a converted World War II minesweeper equipped with large tanks, pumps, and spray equipment traveled south about two miles offshore and sprayed huge clouds of bacteria in a water solution into the air west of San Francisco for 30 minutes each day. On each of the six days, winds blew toward the city and carried the clouds of spray over San Francisco. The paths of the spray and the locations of the sensors collecting the bacteria were shown in Special Report 142. It became my job to analyze the wind flow and the atmospheric stability during each of the tests, labeled, A, B, C, D, E, and F. Tests A, B, C, and D were conducted on September 22 to 25, spraying *Bacillus globigii*, while tests E and F were conducted on September 26 to 27, spraying *Serratia marcescens*. Though there were no known medical problems created by the *Bacillus globigii*, the paths, depth, and progressive densities of the downwind clouds of particles were calculated for all six tests, and the results were compared to the values contained in the 1950 Army Special Report 142.

The thick cloud of *Serratia marcescens* passed the location of the Stanford University Hospital on the evening of September 27 during test F. At the time of the tests in 1950, it was assumed any bacterial infection of individuals affected would be caused by breathing in the bacteria in the cloud and absorbing them in their lungs. However, it was found in the reports of the team of pathologists that investigated the 11 infected patients at the Stanford University Hospital, the bacteria entered their bodies via their urological systems when they were catheterized (as all 11 were there due to urological medical procedures they had undergone).

Test F on September 27 was conducted under very different weather conditions than the first five, as follows:

- There was a rapid temperature increase in height from the surface upward forcing a cap that inhibited dispersion upward of the cloud of particles.
- An abrupt onset of an afternoon sea breeze pushed the concentrated cloud of particles inland and upward over the hills of the city rapidly as they moved inland.
- A wind shift during the test concentrated the cloud towards its north–south midpoint, which passed over the hospital's hill.

The result of these factors was that the concentration of bacteria reaching the Stanford University Hospital on its hill was the greatest of that in any of the six tests.

## DEPOSITION

I had nearly completed the complex analyses of the paths and concentrations of the bacterial cloud when John Kern, the attorney in San Francisco defending the government in the pending suit, called for my deposition to learn what I was likely to testify at trial. He and Ed Nevin flew together from San Francisco to Asheville, and Mr. Kern deposed me at my office on February 26, 1981. He was very well prepared, and his questioning was well organized. He quickly reviewed my qualifications, the data I had relied on, what I had done, and my conclusions—all in a very professional and cordial manner.

## TRIAL

The trial began on March 16, 1981, in the San Francisco Federal District Court, 18 days after my deposition. Federal District Judge Samuel Conti presided, with no jury. The jury box, however, was filled with news reporters, who felt this was a trial likely to be news and worthy of their reporting.

The opening statements by Ed Nevin for the plaintiffs, and John Kern for the defendant, were succinct. Ed Nevin reviewed the essential facts as seen by the plaintiffs:

- the army sprayed the city and the hospital with *Serratia marcescens* bacteria on September 26 and 27, 1950;
- on September 29, infections of patients in the Stanford University Hospital began to appear—the first ever recorded; and

- Edward Nevin, one of the patients, died of a *Serratia marcescens* infection.

His grandson argued the U.S. government was responsible for the untimely death of Edward Nevin. John Kern argued that

- the strains of bacteria that were sprayed and the strain that killed Edward Nevin were not the same, though they had no evidence to support that claim;
- the Army had the "discretionary right" to conduct antibiological warfare tests;
- the Army had tested *Serratia marcescens* bacteria and found them not to be deadly nor seriously infective; and
- the detection of the first hospital case of infection was two days after the tests and therefore unrelated.

Ed Nevin's first witness was Dr. Wheat, who was the senior author of the October 1951 article on the *Serratia marcescens* outbreak at the hospital in 1950. Dr. Wheat testified about the mystery of the 11 cases at the hospital infected by *Serratia marcescens*, which had been considered nonpathogenic, though they discovered in their research for their 1951 article several references to human disease from those bacteria. Dr. Wheat concluded, when he studied the army tests, that there was probability of a causal relationship between them and the senior Edwin Nevin's death.

I was the second witness called by Attorney Nevin and testified about my analysis of the movements and densities of the bacterial clouds created by the army and the fact that the greatest concentrations at the hospital occurred during the final test with the *Serratia marcescens*. Also, when asked, I offered the opinion that the test results could have been obtained from tests over unpopulated areas without exposing the population of a city.

Ed Nevin then called a series of army officials involved with the test, all of whom defended their actions and claimed the bacteria utilized were known to be nonpathogenic and their tests were to determine whether enemy agents could cripple a city by spraying it with dangerous bacteria. All were unaware of Dr. Wheat's article and all felt they were doing their patriotic duty to determine what enemy agents could do in biological warfare.

After the military scientists, Major General William Creasy testified. He had been in charge of the army's biological warfare research at the time of the San Francisco tests. Though retired, he was dressed in full uniform and appeared to consider Ed Nevin as a troublemaker who should not be questioning the army's activities. In the hallway outside the courtroom, the retired general challenged Ed Nevin to a fistfight, which did not occur. In his courtroom testimony, the general contradicted my testimony that the tests could have been conducted over unpopulated land. He claimed the tests could only be done over places where people lived, as the biological warfare was designed to affect people and the test would need to be done where people lived. Though retired, he testified that if he were still in charge he would order the test to be done that same way in the same place. The general was followed on the witness stand by a specialist in infectious diseases who criticized the army tests.

It was natural that Ed Nevin have Edward Nevin, Jr. (his father and the son of the deceased Edward Nevin), testify about Edward Nevin's migration from Ireland to San Francisco, his work for the electric and gas company, his love of the United States, and the family despair on learning that the army of his chosen country may have had responsibility for his untimely death. Likewise, John Kern called several medical experts who believed the bacteria used in the army tests at San Francisco were a different strain than those that infected 10 patients at the hospital and killed the senior Edward Nevin, though they had no such evidence. One expert agreed the tests could just as well have been conducted over an unpopulated area. Several other witnesses testified about hospital record keeping, bacterial mutations, and the likelihood of the bacteria infecting Stanford University Hospital patients being those from the army tests.

## THE VERDICT

Judge Conti ruled that the "discretionary function exclusion" to the Federal Torts Claims Act made the government exempt from legal responsibility in this case because the decision to conduct the tests was made at the level of national planning, rather than at a lower operational level, and was at a sufficient level to exempt the government from liability. While the Nevin family members were disappointed in the verdict, they respected the opportunity to have their day in court.

This was a fascinating and challenging case. Special Report 142 was essential to knowing how, when, and where the bacteria were sprayed into the atmosphere. The weather data, though 20 years old, were well archived at the

National Climatic Data Center, making the detailed wind analysis possible. It was my belief that the test caused the death of Edward Nevin and that there was no need to test the bacteria over a populated area. In 1988, Dr. Leonard A. Cole authored a book titled *Clouds of Secrecy: The Army's Germ Warfare Tests over Populated Areas* (Cole 1988). It describes tests of biological agents conducted in Minneapolis, St. Louis, and the New York subway system and includes a full chapter on the Nevin case.

# CHAPTER THREE

# THE DAY THE SKYWAY FELL

*Failure to Stop and Anchor, Tampa Bay, Florida, 1980*

## THE EVENT

On May 9, 1980, the fierce winds and driving rain striking the U.S. Coast Guard station were diminishing when the radio blared:

> "Mayday, Coast Guard, Mayday. Bridge crossing is down."
> "Mayday, Mayday, Mayday, Coast Guard. Mayday, Mayday, Mayday Coast Guard."

Coast Guard    "Vessel calling Mayday. Vessel in distress. This is United States Coast Guard, St. Petersburg, Florida. Request your position, nature of distress, and number of persons on board. Over."

In the pilot house of the phosphate carrier *Summit Venture*, Tampa Bay pilot John Lerro replied somewhat disjointedly:

Lerro    "This is—all the emergency—all the emergency equipment out to the Skyway Bridge. Vessel just hit the Skyway Bridge. The Skyway Bridge went down. Get the emergency equipment out to the Skyway Bridge. The Skyway Bridge is down. This is Mayday. Emergency situation. Stop the traffic on the Skyway Bridge!"

| Coast Guard | "This is Coast Guard St. Petersburg. Roger. What size is the vessel that hit the bridge? Over." |
| Lerro | "Special. Stop the traffic on the Skyway Bridge. There is some people in the water. Get some emergency equipment out to the Skyway Bridge now!" |
| Coast Guard | "This is Coast Guard St. Petersburg, roger. What vessel are you on? |
| Lerro | "The Summit Venture. The Summit Venture." |
| Coast Guard | "Summit Venture, Coast Guard St. Petersburg, roger. What size is your vessel, and can you assist? Over." |
| Lerro | "Cannot assist. We are six hundred and six feet long, light in ballast. We cannot assist. We hit an abutment. Stop all traffic on the bridge; send some vessels out here to render assistance. People are in the water" (NTSB 1981). |

While this radio traffic was taking place, six cars and a Greyhound bus plunged off the southbound lanes of the Skyway Bridge. The bridge's highest point, where the roadway headed downhill, had fallen into the bay, leaving a huge gap unseen to the drivers cresting the high point. Southbound traffic was stopped when a startled driver realized the roadway was gone, braked to a stop at the very edge of the chasm ahead, backed up to the crest, and blocked both southbound traffic lanes, halting further southbound traffic.

## THE WEATHER

In the early morning of May 9, 1980, the U.S. National Weather Service forecasters preparing the predictions for the Florida Peninsula and offshore waters in the Gulf of Mexico took note of the satellite images of clouds and rainfall echoes from the weather radars at Pensacola (in the western panhandle of Florida), Apalachicola (farther east on the Gulf Coast), and Ruskin (east of Tampa Bay). These all showed a developing squall line over the Gulf waters west of Tampa Bay. The squall line contained thunderstorms with very heavy to intense rainfall. It was moving east at more than 30 miles per hour. Their forecasts called for small craft to exercise caution as the line of thunderstorms passed over coastal areas with winds and seas increasing in the morning hours. These forecasts were being continuously broadcast in the Tampa Bay area on National Oceanic and Atmospheric Administration (NOAA) weather radio.

Scattered showers began west of Tampa Bay shortly after 6:00 a.m. The main thundershower area (in the developing squall line) was seen on radar

50 miles west of Tampa Bay, moving eastward at 32 knots (37 mph) (see Figure 3.1). New, intense thunderstorms were developing offshore west of Tampa Bay and extending south. These were accompanied by an abrupt rise in barometric pressure, a short period of intense rain, and very strong wind squalls with a shift in wind direction from southwest to west-northwest.

## THE SUNSHINE SKYWAY BRIDGE

First built in 1954 for about $22 million, the Sunshine Skyway Bridge consisted of a nearly six-mile-long, two-lane, steel cantilever bridge and causeway that carried U.S. Highway 19 over Tampa Bay and connected St. Petersburg to the north with cities on the west coast of Florida, south of the bay, including Bradenton, Sarasota, Venice, Port Charlotte, Ft. Myers, Naples, and, eventually, Miami on the southeast coast. A twin bridge was completed in 1971 (for around $25 million), increasing the traffic to four lanes with separate bridges for southbound and northbound two-lane roadways.

The Skyway Bridges rose to a high point 150 feet above the water over the shipping channel connecting the many docks of the huge port of Tampa Bay with the open waters of the Gulf of Mexico. At the crest of the rise over the shipping channel, the roadways were made of a steel mesh with small studs to provide traction and reduce the weight of the structure in the 800-foot spans between the tall supporting piers on either side of the shipping channel. The roadways over this span were held in place by huge steel trusses on both sides of the elevated spans. In the early months of 1980, the Skyway had carried approximately 12,000 vehicles per day across the entrance to Tampa Bay, an average of a little more than eight per minute, though in the morning and evening heavier traffic hours, that rate was more than doubled, with vehicles crossing the bridges every few seconds.

## THE SHIP

The *Summit Venture*, a 19,734 ton, 580-foot-long (a few feet shorter than stated in the radio message), 85-foot-wide cargo vessel built in 1976, had sailed from Japan with a cargo of steel destined for Houston, Texas. She then headed, empty, for Tampa to load 28,000 tons of phosphate to carry to Korea. While crossing the Gulf of Mexico, the empty ship was ballasted with seawater. The *Summit Venture* arrived at the approach of the Tampa Bay shipping channel on Tuesday afternoon and anchored for three days, waiting for an open berth at the phosphate docks in Tampa Bay. While at anchor the water ballast was pumped out and the much lighter vessel was readied for her anticipated heavy cargo. The bow was more than 20 feet higher above sea surface than it was with the steel or the water ballast.

## THE CHANNEL

The channel from the Gulf of Mexico into Tampa Bay passes between two islands of the Florida Keys (see Figure 3.2), several miles west of the Skyway in a dredged channel whose edges are marked by numbered buoys. It makes two turns before reaching and passing under the Skyway. Both are turns to the left. Neither is a sharp turn, but each requires meticulous ship handling within the channel to avoid going aground in the shallower water bordering the channel. The second turn is seven-tenths of a mile west of the Skyway and requires precision piloting to aim the ship through the 800-foot-wide opening between the tallest supports of the Skyway.

## FRIDAY MORNING, MAY 9, 1980

The ship was to pick up a port pilot at 5:00 a.m. on Friday morning west of Egmont Key, 11 miles west of the bridge, and enter the bay. At that time, the fog was so thick the captain could not see the bow 500 feet ahead of the pilot house. He requested a delay until the visibility improved. Then, 43 minutes later, the visibility improved as the fog thinned. The *Summit Venture* got underway and headed east toward the channel into Tampa Bay.

The ship's Chinese captain, Hsiung Chu Liu, moved the ship to a position north of Egmont Channel buoy 1, the customary position for taking an inbound pilot on board. While waiting for the pilot, Captain Hsiung Chu Liu noted in the ship's log that the weather consisted of intermittent periods of light rain and drizzle with an estimated visibility of about three miles. At 6:25 a.m., the port pilot Captain John Lerro and an observer pilot in training, Captain Bruce Atkins, boarded the vessel from the pilot boat, which had come along the port (left) side. Atkins was finishing his training to become a member of the Tampa Bay Pilots Association. John Lerro was to conduct the final test on this trip into the bay. Atkins believed this trip to the phosphate loading dock would be the final step of his training period.

Captain Liu briefed them on details of the ship. At 6:30 a.m., Lerro took control and increased speed to half ahead. He then turned control over to Atkins, who headed the vessel into Egmont Channel. There was a radio exchange with a tug and barge, the motor vessel (M/V) *Dixie Progress*, to discuss the weather and their plans for passing each other. Atkins increased the speed to full ahead (13 knots). They entered Mullet Key Channel and passed the M/V *Goodsailor* outbound between the Egmont Lighthouse and buoy 12. A radio communication with the *Pure Oil*, outbound east of the bridge, occurred at 7:15 a.m.

As the ship moved east within the channel at 12 knots (14 mph), bow high in the air, the line of thunderstorms over the Gulf of Mexico, which at 4:00 a.m. was 100 miles offshore, had been racing eastward at nearly 35 mph. At these speeds, the line of thunderstorms would overtake the ship just west of the Skyway at the moment the ship would make the critical second left turn to remain in the channel to pass under the bridge and into Tampa Bay. As the ship passed buoy 14, 3.2 miles west of the bridge, dark clouds, mist changing rapidly to drizzle, and then hard rain driven by strong gusty winds caused John Lerro to take command, relieving Atkins of his duty as pilot. The rapidly worsening weather created a crisis in the pilot house of the *Summit Venture*. The heavy rain became intense; the buoys and the Skyway disappeared from sight in the deluge. Pilot Lerro slowed the ship to a "half speed" of 9.5 knots (11 mph) and ordered the crew to stand by the anchors. The stopping distance of the *Summit Venture* (by backing the engine at full speed) from "half ahead" was 2,520 feet—nearly half a mile. Though the worsening weather, the change of pilots, and the orders to stand by the anchors concerned Captain Liu, he did not question the pilot nor take command.

At approximately the same time, farther west in the channel, Pilot Earl Evans on the vessel M/V *Goodsailor* met the raging squall line head-on. He ordered the vessel to slow to "dead slow," a minimum speed just adequate to maintain steerage. Suddenly the visibility became zero. Thunder and lightning were frequent and intense. Violent gusts of wind, estimated at 60 knots (though neither pilot nor captain looked at the dials of the wind measuring equipment aboard the vessel), blew the pilot and captain into each other on the bridge, and the wind created a frenzy of waves on the water surface. After a brief period the storm swept eastward past the ship and visibility improved. Headed almost into the wind and nearly stopped, the M/V *Goodsailor* weathered the storm without incident and was able to proceed.

The squall line raced toward the stern of the *Summit Venture*, the Skyway, and the *Pure Oil*, piloted by John Schiffmacher. He was aware that the *Summit Venture* was headed east toward him and they would meet near the Skyway. The approaching storm, which he was facing, appeared ominous and he decided to turn out of the channel and anchor until both the squall line and the *Summit Venture* passed and he could have good visibility to pass under the Skyway. He ordered a turn to the right out of the westbound channel. The rain became intense and blew by nearly horizontally. The radar screen was "saturated" by the strong rain echoes (images of varying colors indicating the intensity of rainfall), and all navigational features were obscured by the rain echoes on the screen.

The line of thunderstorms overtook the *Summit Venture* from behind at 7:23 a.m. as the vessel passed between buoys 15 and 16 (two miles west of the bridge), headed for the critical turn between buoys 1A and 2A, just seven-tenths of a mile west of the opening under the Skyway. The increasing winds from the southwest, which had caused the vessel to move slightly left of its heading, suddenly became violent wind squalls from the west-northwest, blowing from behind and against the port (left) side of the *Summit Venture*, pushing the vessel with far greater force to the right of its heading. Bruce Atkins had been studying the radar, watching for the images of the buoys 1A and 2A, knowing the ship must pass between them to make a 16° turn to the left in order to "shoot the notch"—the 800-foot-wide channel beneath the high arch of the highway. Just before the torrential rains hit the ship, Atkins had a brief glimpse of the buoy images on the radar and told Lerro they were in the channel. Then the rain echoes blanked out all returns and the radar became useless. The bow of the ship disappeared in the intense rain.

At 7:28 a.m. a lookout on the bow of the *Summit Venture* reported seeing the buoy on the port side. The color, shape, and presence or absence of sound signals from the buoy were not discussed. He assumed it was buoy 1A on the left side of the channel and ordered a turn to the new heading of 063 degrees, which, under normal conditions would have taken the ship in the channel through the Skyway opening. Conditions were far from normal, and the confusion over the location and identity of the buoy delayed the order to turn.

In fact the buoy to the left of the ship was not 1A but was 2A, which marked the right side of the inbound channel. The ship was outside the channel far to the right of its intended course. The turn would normally have begun before reaching the buoys 1A and 2A, placing the ship midchannel on the new heading as it passed between these critical "turning buoys." The mistake on the identity and location of the buoy and the delay in making the turn aimed the *Summit Venture* at the bridge far to the right of the ship channel, with the radar useless and nothing visible to those on board but blinding wind-whipped rain. The ship continued blindly forward.

Lerro became startled by the force of the wind and became aware of the *Summit Venture*'s crablike sideways motion because of the shifted winds. He told the captain to order the crew to stand by the anchors in case the ship was blown out of the channel. He ordered the engines slowed to "slow ahead" 6.2 knots (7 mph) and continued to peer ahead, searching for the opening. In a brief instant, a lessening of the blinding rain between squalls gave him a glimpse of the vague outline of the vertical supports and steel girders of the Skyway much too close to bring the ship to a stop. "He ordered left full rudder, a double full astern bell, and the anchors let go" (Grace 1981, p. 15). At that instant, the *Summit Venture* struck a vertical supporting column of the Skyway 200 feet to the right of the channel. The column was designed only to support the Skyway and was not designed to resist a horizontal force. It collapsed; 1,260 feet of the southbound lanes of the Skyway fell. From the midpoint of the grillwork over the channel southward for nearly a quarter of a mile over two supporting columns, the roadway fell into Tampa Bay. A 30-yard segment of it fell onto the bow of the *Summit Venture* and remained there. Southbound traffic plunged into the bay; 35 people lost their lives, 21 of whom were on a Greyhound bus.

As Pilot Schiffmacher on the *Pure Oil* prepared to anchor, the storm moved east and the weather cleared rapidly. He was about to decide not to anchor but to simply complete a slow circle and then continue westbound toward the Skyway into the clearing weather. At that moment he and the captain heard John Lerro's voice with his urgent "Mayday, Coast Guard, Mayday . . ." message. Looking west, he and the captain were aghast at the sight of the Skyway with a huge segment of the southbound lanes missing.

In the following several days, a total of 35 bodies were recovered from the waters of Tampa Bay. The single occupant of a pickup truck rode a small section of the 1,280 feet of falling highway that landed on the bow of the *Summit Venture*. After the truck bounced and fell into the bay, the driver was able to swim to the surface and was pulled aboard the ship. The companion bridge carrying the northbound lanes was slightly damaged but declared safe by inspecting engineers. Its traffic lanes were repainted with double yellow lines and two-way traffic proceeded after a few days. Many sightseers used it as a vantage point to view the fallen southbound Skyway.

All ship traffic was halted by the Coast Guard from May 9 to 13, while the steel girders that were blocking the shipping lanes were cleared. On May 13, an auxiliary channel for shallow draft (small) vessels—with tug assistance—was opened north of the main channel. Then, 10 days after the collision, half of the main channel was opened to one-way daytime use, with tug assistance; 45 days later, the main channel reopened. The delays in shipping to and from the Port of Tampa created tremendous financial losses to businesses and shipping companies.

## LITIGATION

The following lawsuits were immediately filed:

- The state of Florida filed a suit against Hercules Carriers, Ltd., the owners of the *Summit Venture*, registered in Monrovia, Liberia, and operated out of Hong Kong by a Chinese crew, to recover the costs related to the loss of the Skyway.
- A marine board of investigation was convened to review the actions of Pilot Lerro.
- The owners of the *Summit Venture* filed a suit to be absolved from blame or to have their liability limited to the $14 million value of the vessel (a practice having its origins in British maritime law).
- Relatives of several passengers on the Greyhound bus filed a suit against Greyhound Bus Lines, alleging the driver recklessly drove the bus at excessive speed past persons attempting to signal the bus to stop.
- The mother of a college student from Tallahassee, Florida, on the Greyhound bus filed a $6 million suit against the deceased driver of the bus for the death of her daughter.
- Commercial firms and shipping interests filed suits against the state of Florida and the owners of the *Summit Venture* to recover their losses because of the closing of the shipping channel.
- Personal representatives of vehicle passengers killed in the accident brought suit against the Department of Commerce, alleging the National Weather Service did not properly monitor or warn the public of the weather conditions.

The owners of the *Summit Venture* and the complex of interlocking Hong Kong corporations operating the ship claimed the collision was due to an "act of God," which produced the unprecedented unforeseeable, overwhelming weather causing the accident. They were represented by Tampa attorney Dewey R. Villareal, Jr. He was a graduate of the U.S. Coast Guard Academy, Tulane University, and the University of Florida Law School. He had been a professor of admiralty (maritime) law at Stetson University and was a senior partner of the law firm of Fowler, White, Gillen, Boggs, Villareal, and Banker in Tampa.

A number of claimants against the ship and its owners, who believed the ship owners were responsible because their agent (the captain) allowed Pilot Lerro to continue toward the bridge when visibility in the rain was less than 500 feet (maritime rules require that vessels be able to stop within half the distance mariners can see ahead of them), were represented by Roger A. Vaughan, Jr. He was a graduate of the U.S. Merchant Marine Academy and George Washington University School of Law and an experienced admiralty lawyer. The state of Florida was both a plaintiff in suits against the owners of the *Summit Venture* and a defendant in several suits alleging negligence in failing to protect the public against just such a disaster.

The Tampa-based law firm of Shackleford, Farrior, Stallings, and Evans represented the state of Florida in the major suits for and against the state. That firm's principal trial attorney in the suits involving the collision was David Hanlon, a graduate of the U.S. Merchant Marine Academy, Duke University, and Southern Methodist University (where he received his law degree). These three highly qualified and experienced admiralty attorneys respected each other and were widely known for their professional conduct.

David Hanlon realized the importance of a detailed analysis of the weather in the pending litigation and asked his law partner, Benjamin Hill, to find a weather consultant who could reconstruct the weather events of the day and testify about these in the various suits and hearings. Attorney Hill contacted the American Meteorological Society and obtained a list of board-certified consulting meteorologists. He also contacted major law firms specializing in marine accidents to obtain information on weather witnesses they had used. In comparing both lists he found the Climatological Consulting Corporation. He traveled to Asheville to interview me and my associate Henry Harrison (former Superintendent of Meteorology at United Airlines) shortly after the event. We were ultimately engaged by the law firm on behalf of the state of Florida.

Key weather-related questions in the Skyway collapse litigation included:

- a precise reconstruction of the sequence of weather conditions affecting the shipping channel to and through the Skyway on the morning of May 9, 1980;
- a description of the forecasts of the weather of that time and place as issued by the U.S. government;
- when and how individuals could obtain information regarding the weather in existence and forecast; and
- what individuals on vessels eastbound or westbound in the channel could have perceived with their own senses and/or with the aid of technical devices—such as radars in the pilot houses of the vessels—about the weather occurring and about to occur in their vicinity at successive intervals of time.

The answers to these questions could be relied upon by other experts to testify whether the actions of those on the scene at the time were appropriate in view of the weather. The answers could be derived from an analysis of the following:

- Satellite data—in the form of visible light photographs of clouds and infrared images of the temperatures of the top of those clouds (and surrounding land and water surfaces)—were available, usually about 12 minutes after being measured at 30-minute intervals.
- Weather radar images of the intensity of the return energy of the horizontal beam reflected by falling rain drops—mostly at five-minute intervals—by weather surveillance radars at Pensacola, Apalachicola, and Ruskin, Florida, were accessible. Their information in 1980 was not as readily available to the public and special interest groups as it is today, but it was very valuable in the detailed reconstruction of what occurred. (The data from the nearest National Weather Service radar—at Ruskin, Florida, east of Tampa Bay—were interrupted at the time of the collision because of a lightning strike on the radar transmitter/receiver site, about 15 minutes prior to the collision and did not resume until 30 minutes later.)
- On the other hand, each of the vessels in the channel was equipped with operating radar that produced continuous (though not recorded) images of the rainfall intensity within several miles of the ships, until each vessel was within an area of such heavy rain that the radar returns from the rainfall "saturated" the radar screen and obscured (wiped out or overwhelmed) all other echoes, such as other ships, buoys, and the bridge structure.
- The official weather observations, made by the National Weather Service at hourly intervals mostly at airports; the images of the weather satellite (both visible and infrared); the weather surveillance radar images; and the multiplicity of weather forecasts made for public dissemination, aviation interest, mariners, and so on, were available (all of which were stored at the National Climatic Center, now the National Climatic Data Center).
- Copies of the government-analyzed weather maps at time intervals of three hours (eight analyses per day) are similarly archived and available. The 7:00 a.m. eastern standard time (EST) [8:00 a.m. eastern daylight time (EDT)] weather maps from data observed 30 minutes after the ship/bridge collision are published by the government and available to any user.

- The eastward progress of the squall line was clearly shown on the 2:00, 5:00, and 8:00 a.m. (EDT) analyzed surface weather charts on May 9, 1980, as it intensified while approaching and crossing Tampa Bay on Florida's west coast.
- Not archived, but reconstructed from the basic data, the content of the weather message continuously broadcast via NOAA weather radio, a common source of coastal weather data, information and forecast relied upon by mariners, was available.
- Local data from the pilot station on Egmont Key, and from the several ships entering or leaving the port of Tampa, were not archived. They were, however, essential for the reconstruction of the local details and timing of the most critical and rapidly changing weather of the line of thunderstorms creating the blinding rain and wind squalls that swept across the ship channel and bridge that morning. These were obtained from the eyewitness testimony of individuals given in their statements and depositions after the event.

In preparation for testimony at the various hearings, depositions, and trials, the forensic meteorologists not only collected all the archived data but attended depositions and hearings where individuals on the scene testified about the weather they experienced on the fateful morning of May 9, 1980. The attorneys for the pilot, the shipping company, and several plaintiffs hired Bill Kowal, a local Tampa meteorologist who worked both for a television station and the forensic meteorological firm in Tampa, Gulf States Weather. Both CCC and Gulf States Weather utilized the same basic data and information. Though the reanalysis of the weather events were quite similar, the interpretations of what the individuals involved in the accident would have known as events evolved differed—a common occurrence in expert testimony.

Each forensic meteorologist was first engaged as a consultant to confer with the hiring attorney to describe their reconstructed weather analysis and discuss their potential testimony. None of their work as a consultant could be "discovered" by the opposing attorneys. When the lawyers were satisfied that their consultant's potential testimony would be useful, they named them as potential witnesses. All work after that was subject to discovery (either by copy of any reports they made or through questioning in depositions). To assure full understanding of all aspects of their potential testimony, the engaging attorneys provided essential information on the likely topics to be placed before the courts and held several meetings with their named witnesses to review weather-related details and coordinate the content and preparation of illustrative visuals to be used as exhibits at trial.

The Shackleford firm in Tampa, representing the state of Florida, assigned a young law clerk, David Stefany (now a senior labor lawyer), to maintain detailed records on all correspondence, reports, discussions, and so on, with me, their weather expert; keep the firm members and the expert fully advised on developments in trial preparation; and coordinate attendance at all pertinent depositions or pretrial conferences. That liaison (unusual, but most welcome) was extremely beneficial in preventing any gaps in mutual awareness of significant trial-related events. Many of the defendants in these various suits claimed an "act of God defense."

Among the early legal proceedings was a U.S. Coast Guard marine board of investigation (Grace 1981, 27 pp.), which sought to investigate all the available facts and develop conclusions as to "the proximate cause of the casualty." They concluded the fault was "the decisions of the pilot, John E. Lerro to continue to attempt to pass under the Sunshine Skyway from the vicinity of the intersection of Mullet Key Channel and cut A channel" (Grace 1981, p. 23) (marked by buoys 1A and 2A). They also found that the captain of the *Summit Venture* was at fault because "he did not fully exercise his responsibilities for the safe conduct of his vessel in that he expressed no concern to the pilot when he did, in fact, have misgivings. . . . [and that] he abdicated his role [as captain] and relied entirely on the abilities and judgments of the pilot. . . ." (Grace 1981, p. 24). Soon thereafter, a special panel of the state Board of Pilot Commissioners held hearings. They concluded: "It was Pilot John Lerro's error that caused the Sunshine Skyway Bridge to tumble into Tampa Bay" (Curtis 1980, p. 37). Lerro's defense attorneys vehemently argued the "act of God defense." Bill Kowal, the meteorologist from the Tampa television station and Gulf States Weather, testified that the weather was in fact so overwhelming, unprecedented, and unforeseeable that it became a "force majeure" and caused the accident. His testimony was augmented by a naval architect who indicated the sudden weather changes forced the vessel to move far to the right of the steered course (an opinion in conflict with the calculations of the U.S. Coast Guard), which showed the late turn after the confusion over the buoy put the ship far to the right of its proper course.

I, on the other hand, testified with the aid of several illustrative visuals mounted on large poster boards that the squall line was well predicted by the National Weather Service, that it was prominently mentioned on the frequently updated messages broadcast over NOAA weather radio, that these could be heard in the pilot house of the *Summit Venture*, that the radar on the *Summit Venture* would have shown the intensity of the rain rapidly overtaking the ship from behind, and that an occasional look back over their shoulder would have alerted the pilots, captain, mate, or crew of the severity of the approaching ominous weather in ample time to anchor the vessel before the deluge associated with the squall line reduced visibility aboard the ship to near zero. In March of 1981, however, the Board of Pilot Commissioners reversed their earlier findings. They voted unanimously to restore John Lerro's license based on their new ruling that "he acted in a reasonable [and] prudent manner on May 9, 1980" (Curtis 1980, p. 40). A few weeks later, a greatly divided National Transportation Safety Board issued a report in which two of the four members faulted John Lerro for not stopping or turning aside when the storm struck, one member dissented and absolved the pilot of responsibility, and one blamed the National Weather Service for an inadequacy of warnings. Their findings, however, carried no authority with the Florida Board of Pilot Commissioners, and Lerro returned to duty as a Tampa Bay pilot in April 1981.

No court sustained the allegations against the National Weather Service for inadequate warnings. Indeed, the weather information available in the coastal forecasts and continuously broadcast over NOAA weather radio made prominent mention of the fast-moving squall line threatening the Tampa Bay area well in advance of the ship–bridge collision. The owners of the *Summit Venture* filed suit in Federal District Court for the Middle District of Florida "seeking exoneration from or limitation of liability" shortly after the collision. U.S. District Judge George C. Carr held preliminary hearings to determine "whether John Lerro, the compulsory pilot aboard the *Summit Venture*, was in any way at fault for the vessel's collision with the Skyway Bridge" (Carr 1982a, p. 2). Testimony accepted by the court included the following:

- "When visual visibility is reduced below one half mile (approximately 2,600 feet) [in ship channel or near obstacles], a vessel should be anchored [until the visibility improved]" (Carr 1982a, p. 4).
- ". . . the speed of the vessel in restricted visibility should be reduced so that the vessel can be stopped in one half the distance of her visibility" (Carr 1982a, p. 4).
- "Navigation solely by reference to radar is improper" (Carr 1982a, p. 4).

The court found—based on the facts and testimony presented—that these three conditions existed by the time the *Summit Venture* reached buoy 16, two miles west of the Skyway, and that the ship should have been anchored at that point until the weather cleared. Had the ship stopped at that point, further deterioration of the weather would not have been a cause for collision. The court ruled that "no real dangers existed or were inherent in the weather itself, the weather merely set the stage for Lerro's original negligence (in failing to stop and anchor in the area of buoy 16); the weather was not a legal proximate cause, but at best a remote cause or condition [of the collision] . . . Lerro's failure to stop and anchor was active negligence. Whatever happened from that point on was the direct and proximate result of Lerro's failure to anchor and led directly in an unbroken sequence to the tragic collision that occurred some ten minutes later" (Carr 1982b, p. 29).

Judge Carr denied Hercules Carriers exoneration from or limitation of liability on July 2, 1982. Later, the same court, with Senior District Judge Daniel Thomas presiding, held hearings beginning in October 1982. The judge heard evidence, examined exhibits, considered pleading stipulations, and proposed findings of fact and conclusions of law offered by the attorneys for Hercules Carriers, several wrongful death and personal injury claimants, the state of Florida, Greyhound Bus Lines, and the U.S. admiralty and shipping section of the U.S. Department of Justice. He affirmed the prior decision of Judge Carr; he found the captain of the ship negligent for not taking over control of the vessel when he felt concerns over Lerro's continuing toward the bridge when visibility became less than the stopping distance of the vessel and Hercules Carriers negligent in not properly training the officers and crew of the full responsibilities of deck officers, lookouts, anchoring personnel, and a number of other duties. There were many other legal decisions, including the damage trials, in which the state of Florida was awarded recovery of more than $20 million for the collapsed structure. In all, the courts ruled that while the weather was a contributing factor, it was the negligence of individuals in not properly recognizing and responding to the changing weather conditions that caused the tragedy.

Only days after the collision, single-lane, two-way traffic was routed across Tampa Bay on the previously northbound lanes of the remaining eastern spans of the prior twin bridges, and construction began on the seven-year, $244 million project of building a new Sunshine Skyway Bridge designed to overcome the safety concerns of the original bridges. The center span over the ship channel was increased 50% in width from 800 to 1,200 feet. The clearance from the water level to the center span base was raised from 150 to 190 feet, allowing taller modern ships greater clearance. More than 300 precast concrete segments were linked together with high strength steel cables to form the 29,040-foot-long (5.5 mile) pair of 40-foot-wide roadways. Held up in the center by 21 huge steel cables, each encased within nine-inch diameter steel pipes painted bright yellow (representing the "Sunshine State"), the roadways provide unobstructed views of the water beyond the low road edge barriers. The six major supporting piers are surrounded by huge concrete islands—known as dolphins—each designed to withstand the force of the impact of a collision by an 87,000 ton ship (nearly 4.5 times that of the *Summit Venture*). The new Skyway opened to traffic in 1987 as the world's longest cable stayed concrete bridge (PBS Online 2001). The older bridges were demolished except for mile-long, low segments at the northern and southern shores of the bay, which remain as major sport fishing piers.

Though I had watched the 7:00 a.m. news broadcasts on national television that featured the satellite and radar imagery showing a huge comma-shaped storm over the northeastern Gulf of Mexico with the tail of the comma headed east toward Tampa Bay, I had no idea I would be involved in such a high-profile case. The weather analysis was straightforward, but the legal complexities of the multiple suits were overwhelming. In my opinion, John Lerro had all the weather information available to him to avoid this disaster but ignored it at the very high cost of many lives.

# ALICIA AND TANK 089

*Revealing Oil Stains, Galveston, Texas, 1983*

## THE EVENT

The winds of Hurricane Alicia severely damaged Tank 089 at the Texas City Refinery near Galveston, Texas, during the early morning darkness of August 18, 1983. Tank 089, which was 340 feet in diameter and 48 feet tall with a capacity of 740,000 barrels of oil, was the largest of the more than 100 storage tanks at the refinery. It was believed to be the largest tank in the world at that time. The floating circular top of the tank was crumpled and sank, a 12-foot section near the top of the north wall was buckled, fittings near the base of the tank were damaged causing the stored oil to drain out of the tank, and the long, spiral ladder on the outside of the tank, providing access to the top, was bowed and twisted.

## HURRICANE ALICIA

Alicia, the first tropical storm of the 1983 hurricane season, developed on August 14 about 200 miles off the shoreline in the western Gulf of Mexico. During the next 24 hours, the storm intensified rather rapidly. The central pressure dropped steadily for the next two days as the storm grew in size and strength and drifted first westward and then northwestward toward the Texas coast (see Figure 4.1). Tropical Storm Alicia became a hurricane when the winds near the storm center reached 75 mph on August 16. It continued to strengthen to a minimal category 3 on the Saffir–Simpson hurricane scale (winds 111–129 mph) on August 17. The central eye of Hurricane Alicia made landfall just west of Galveston Island at about 1:30 a.m. [central standard time (CST)] on August 18 and then proceeded northward over and past Houston. Alicia, unlike most hurricanes, did not weaken rapidly after making landfall but continued above hurricane strength until passing well north of Houston and northbound into Oklahoma where it turned northeastward and gradually weakened as it approached Nebraska. Hurricane Alicia produced 22 tornadoes: 20 of these were rated weak and short-lived at F0 (wind speeds of 40 to 72 mph), one at 1:15 a.m. (CST) on August 19 was believed to be F1 (wind speeds of 73 to 112 mph), and one, nearly 150 miles inland, was believed to be F2 (wind speeds of 113 to 157 mph). None were near the refinery's tank farm.

The maximum winds of Alicia dramatically impacted the cities of Galveston and Houston. Though Alicia was, by virtue of its wind strength, a minimal category 3 storm on the Saffir–Simpson scale, it caused more than $2 billion worth of total damage in this area. The tremendous damage was largely because the area of maximum winds crossed large metropolitan areas, placing a huge network of expensive structures, buildings, and urban infrastructures at risk. Windblown debris and roof gravel proved devastatingly damaging to the glass of Houston's skyscrapers.

With no electricity or water, downed power lines and trees, glass shards falling onto the streets and sidewalks, and battered, windowless skyscrapers, the central business districts of the two major cities of Galveston and Houston were cordoned off and vacated for several days. The extensive damage to the Galveston–Houston area—both of which were described as being left a shambles—caused the American Society of Civil Engineers, the National Academies of Sciences and Engineering, and at least half a dozen professional organizations and agencies to convene a special conference on August 16–17, 1984, in Galveston, Texas. *Hurricane Alicia: One Year Later* (Kareem 1985) featured 29 learned papers on meteorology, structural behavior, window glass and curtain walls, testing for hurricane-resistant design, (building) codes and standards, and designing for the future. The proceedings were published by the American Society of Civil Engineers. By being within radar range of the National Weather Service radar at Galveston and by being nearly continuously observed by aircraft reconnaissance, as well as by directly affecting a large number of weather observing sites in the heavily populated and industrial areas of coastal Texas, Alicia was one of the most comprehensively observed hurricanes ever to affect the United States both prior to the time of striking the Texas coast and for many miles inland.

## THE LITIGATION

The huge, cylindrical Tank 089 with its massive floating top was built in 1981. It had been designed to withstand winds of 125 mph. The refinery owners believed the winds at that location were less than 125 mph and that the structure failure was due to faulty design or construction. They filed suit against the tank manufacturer for the cost of the damages. (Though other tanks were damaged by the wind, they were not part of this litigation.) The tank manufacturer claimed the tank had lived up to its design standards and must have been damaged by winds in excess of 125 mph, either directly from the winds of Alicia or from a tornado embedded within the hurricane. Cooper Ashley, the attorney for the refinery, initially engaged a college professor teaching meteorology at a university in Houston. The professor prepared a detailed wind analysis, based largely on the information published in the proceedings of the symposium held one year after Alicia's onslaught. He augmented the symposium proceedings with original data—including recorded wind speeds—from four locations near (but not on) the Texas City Refinery and from wind data continuously recorded aboard the U.S. Coast Guard vessel *Buttonwood* moored at the Galveston Coast Guard station in Galveston Bay within a few miles of Tank 089.

Nearly six years after the event, the meteorology professor was deposed on August 10, 1989. He testified that "it is my opinion the sustained winds [of Alicia at Tank 089] never exceeded 125 mph." When asked "What about gusts?" he replied, "There may have been gusts just barely over 125 mph—probably no more than 130 mph" (Meisner 1989, p 34). To counter the plaintiff's expert's testimony, James Ansell, the attorney for the defense (tank manufacturer) engaged a forecaster/forensic meteorologist from a private weather firm in Pennsylvania. He made a detailed study of the winds that impacted the tank and was offered as an expert. His opinion was that the weather circumstances were such that they could have produced winds of 130 mph. With both witnesses unable to provide a definitive maximum wind speed at the tank, Mr. Ashley sought further expertise. He contacted Dr. Tetsuya "Ted" Fujita, who was willing to perform the analysis but not willing to testify. Ted recommended Mr. Ashley hire Climatological Consulting Corporation, which he did in January of 1990.

Dr. Fujita and I had become professionally and personally acquainted while attending scientific meteorological conferences. He was a renowned meteorologist at the University of Chicago, where he headed the Wind Research Laboratory. Specializing in damage-producing wind events, particularly tornadoes and hurricanes, he published landmark research papers that led to his becoming known as "Mr. Tornado." He also became known as a great meteorological detective with powerful intuitive skills, able to formulate conceptual models based on complex clues left behind by nature at the disaster scene.

At the request of Mr. Ashley, Dr. Fujita and I performed a detailed analysis of the wind associated with Hurricane Alicia and its impact on the Texas City Refinery and specifically Tank 089. In doing so, we looked at the entire dynamics of the hurricane, but we focused on wind and addressed the specific questions he asked us to address: Was there or was there not a tornado that traversed Texas City Refinery? How strong were the winds at Tank 089? What direction did they come from and when did they occur?

After Dr. Fujita and I visited the site of Tank 089 on March 3, 1990, we reviewed all prior studies and reports, analyzed the recorded data, visited the data recording sites, viewed the photographs of the sequential radar images of the precipitation as recorded from the National Weather Service surveillance radar at Galveston, Texas, read all the eyewitness accounts of the events in the Galveston area during Alicia, and viewed hundreds of photographs taken soon after the storm. Extremely detailed analysis of the winds of Alicia utilizing all the available information from 17 anemometers (ground-based wind measuring devices), multiple aircraft flights at 5,000 feet above ground level, and precipitation echo movements observed by radar showed that the maximum winds of the storm over land were 90 mph by radar, 100 mph by anemometer, and 115 mph by aircraft measurement at 5,000 feet. The strongest peak gust measured was 105 mph from the southeast at 1:30 a.m. on August 18 by the U.S. Coast Guard Cutter *Buttonwood* moored in Galveston Bay whose anemometer was 45 feet above the water. The wind directions and speed data from the four recording anemometers nearest the tank farm were interrupted by a series of power failures during this storm. Based on our detailed analysis of the winds, a review of the reported tornadoes, investigation of the passage of rainbands across the area, the radar photographs, and the available wind trace recordings from the four anemometers in the close locale, we ruled out the possibility of any tornadic patterns in the vicinity of Tank 089. Without any evidence of tornado damage, we set out to determine the wind speed at the time the tank ruptured.

The photographs taken in the tank farm by employees immediately after the hurricane from several angles were of special value in determining the time the tank ruptured. They showed oil stains on the upper portions of the outside, west, northwest, north, and northeast sides of the tank, but none on the east through southwest sides. These stains had to have been made by oil spilling onto the downwind sides of the tank after the top ruptured and as the hurricane center moved north past the west side of the tank. The sequence of hurricane winds experienced by the tank came—in time sequence—from the east, then southeast, south, and southwest (see Figure 4.2). Analysis of the timing of the sequence of wind direction and speed made it possible to determine the time of initial failure when the east winds first drove oil over the western rim of the tank.

Dr. Fujita pointed out that the filmed radar images on the National Weather Service Weather radar at Galveston provided direct indisputable evidence of the wind speed at Tank 089 continuously during the storm. The radar images of the rain showers in Alicia were photographed every 40 seconds throughout the storm's passage. Tank 089 reflected the radar beam and showed brightly on the images as a stationary bright target in each photograph, while the images of the rainfall, blown by the wind, moved with the speed of the wind from image to image. By plotting the sequential rainfall images and measuring their movement in 40-second intervals between frames, we could determine the speed of the wind moving the showers past the tank. They were compared with the wind data (when available) from the recording sites and found to be in agreement. These detailed analyses showed that winds from the east blowing at 85 mph impacted the tank at 11:30 p.m. on August 17 and that (maximum) winds from the southeast struck the tank at 4:30 a.m. on August 18 (five hours later), blowing at 90 mph with gusts of 110 to 115 mph. Tank 089 had failed in winds of less than 125 mph, proving faulty construction. It was only through Dr. Fujita's clear insight that we were able to solve this mystery.

## THE TRIAL

At trial, I was able to demonstrate to the jury the motions of the rain images within Alicia's rainbands in relation to the stationary tank image by utilizing a time-lapse projector to show the successive images on a backdrop of the areas of coverage of the Galveston-based radar, which included the Texas City Refinery and Tank 089. The jury was able to see the moving rainbands coming with increasing speed and changing direction past the stationary image of the tank during the night. The visual evidence coupled with the stain patterns convinced the jury of the early failure with winds of only 85 mph. They awarded the refinery $2.5 million for the damages.

Even though this hurricane was so well documented, it still required the brilliant mind of Dr. Fujita to solve this case. With his amazing ability to "think outside the box" he used the oil stain details on the side of Tank 089 together with the radar images and wind data to demonstrate the failure in winds lower than the tank was designed to withstand.

# CHAPTER FIVE

# DELTA 191

*"Lightning Out of That One," Dallas/Fort Worth, 1985*

## THE EVENT

One hundred and twenty-six passengers, eight crew members, and one car driver on the ground were killed when Delta Air Lines flight 191 crashed on final approach to land at the Dallas/Fort Worth International Airport (DFW) at 6:05 p.m. (CDT) on August 2, 1985. Twenty-six passengers and three crew members survived.

The half-filled, 302-seat Lockheed 1011 Tristar aircraft was bound from Fort Lauderdale, Florida, to Los Angeles, California, with a scheduled en route stop at Dallas/Fort Worth, Texas (Figure 5.1). In the final stages of its descent to land, the plane encountered a microburst in the rain beneath a rapidly developing thunderstorm one mile north of the approach end of runway 17L at the airport. The aircraft was driven to the ground, struck a car on the service road north of the airport, slid onto the field beside the runway, struck two huge water storage tanks, and broke apart and burst into flames, which consumed the front section. The 26 survivors were seated in the rear section, which broke away from the main body and escaped the fire as the two parts of the plane slid across the ground east of the runway.

Fire trucks were dispatched to the scene within 45 seconds, and emergency equipment was called on from adjacent communities, though some were delayed by communication problems. The survivors were transported to

41

hospitals in Dallas and nearby communities. The "Go Team" of the National Transportation Safety Board in Washington, D.C., rushed to the scene overnight and began their investigation of the crash. Hundreds of other investigators, relatives, insurers, lawyers, and concerned individuals began gathering information, both to assist the NTSB and prepare for the inevitable litigation.

## HISTORY OF THE FLIGHT

The plane had departed from Fort Lauderdale at 4:10 p.m. (EDT) [3:10 p.m. (CDT)], 3 hours and 55 minutes earlier in good weather and climbed uneventfully to a series of cruising altitudes above 30,000 feet. According to protocol, the Delta Air Lines Dispatch Office in Atlanta had electronically provided the flight crew a document package that included pertinent information on all aspects of the flight, including detailed weather information and forecasts for all segments of the route. It stated, in part, that there was a possibility of widely scattered showers and thunderstorms becoming isolated after 3:00 p.m. (CDT) over the western portion of the route. In particular, "Metro Alert No. T87," valid until 9:00 p.m. (CDT), stated that "an area of isolated thunderstorms is expected over Oklahoma and northern and northeastern Texas . . . a few isolated [cloud] tops to above FL 450 [45,000 feet]" (NTSB 1986, p. 1).

The captain made a routine position report to the Delta dispatch office when passing over New Orleans. He was aware of a line of showers and tall clouds intensifying along Louisiana and the Gulf Coast ahead of the flight along their projected route to an approach point named "Scurry," 35 miles southeast of DFW. To avoid that developing weather he requested that Air Route Traffic Control (ARTC), which separated planes along air routes, change their flight path farther inland to approach DFW from a navigational point named "Blue Ridge," 35 miles northeast of DFW (Figure 5.2). The controllers advised that it would require a delay because of heavy traffic. The captain replied he'd "'rather wait for an airplane than, uh, go fight a bunch of weather' [the weather southeast of DFW]" (Belew 1989a p. 21). The flight was rerouted and required to fly a holding pattern for 10 minutes before being merged into the stream of air traffic approaching DFW via the Blue Ridge approach.

After Blue Ridge, the flight would be handed off to a series of air traffic controllers known as "Feeder East" and then a terminal radar control (TRACON) controller, who would direct a series of planes one minute (three miles) apart through a sequence of descending altitudes to a position five miles north of the airport at an electronic "outer marker" where a local controller in the tower at DFW would take over control along the final descent path to the runways. While approaching Blue Ridge at 5:35 p.m. (CDT), the flight crew received the automated information service (ATIS) transmission from 4:47 p.m., which indicated the weather at DFW was scattered (less than half the sky covered) clouds at 6,000 and 21,000 feet, temperature of 101°F, dewpoint of 67°F, wind calm, altimeter setting of 29.92", runways 18 right and 17 left in use for landings, and visual approaches in use. Still in radio contact with the Fort Worth Air Route Traffic Control, the captain requested a deviation around a thunderstorm near the Blue Ridge intersection at 5:44 p.m. Though the hour-old ATIS indicated good flying weather at DFW, clouds were building a short distance north and northeast of the airport.

At 5:49 p.m., a radar image of a developing "cell" a few miles north of DFW appeared on both the weather radar at Oklahoma City, Oklahoma, and at Stephenville, Texas (72 miles southwest of DFW). The cell intensity on both radars was "Level 1." To communicate to the pilots the intensity of a thunderstorm, weather radar operators use Video Integrator Processor (VIP) levels. "The National Weather Service (NWS) radar observer is able to objectively determine storm intensity levels with VIP equipment. These radar echo intensity levels are on a scale of one to six. If the maximum VIP Levels are 1 "weak" and 2 "moderate," then light to moderate turbulence is possible with lightning. VIP Level 3 is "strong" and severe turbulence is possible with lightning. VIP Level 4 is "very strong" and severe turbulence is likely with lightning. VIP Level 5 is "intense" with severe turbulence, lightning, hail likely, and organized surface wind gusts. VIP Level 6 is "extreme" with severe turbulence, lightning, large hail, extensive surface wind gusts and turbulence" (FAA 1983, p. 5). The VIP Level 1 indicated light rain. No mention of this new rain cell was made to pilots of approaching flights. At 5:51 p.m., the National Weather Service observer at DFW completed his routine hourly weather observation. In remarks at the end of the observation, he noted the entire sky north and northeast of the airport was blanketed by cumulonimbus clouds.* At that time Delta 191 was descending through

---

* "Cumulonimbus: (Abbreviated Cb.) A principal cloud type (cloud genus), exceptionally dense and vertically developed, occurring either as isolated clouds or as a line or wall of clouds with separated upper portions.

These clouds appear as mountains or huge towers, at least a part of the upper portions of which is usually smooth, fibrous, or striated, and almost flattened as it approaches the tropopause" (AMS 2012a).

11,000 feet over the Blue Ridge locator 35 miles northeast of DFW, heading toward the airport well north of the building clouds near the field.

Four minutes later, a local controller in the tower cab saw a cloud to ground lightning bolt east-northeast of the field from the clouds developing there but did not notify anyone. Three mechanics on the ground also noticed lightning coming from the same developing storm, which they, later, said was producing the heaviest rain they had seen during their nine years at DFW. They did not notify anyone at the time. By 5:54 p.m., when Delta 191 was 18 miles north of the field, later analysis showed that the National Weather Service radar at Stephenville indicated the light rain shower that had been a Level 1 echo was growing quite rapidly north of the field and had enlarged to five or six times its former size. The radar operator in Stephenville was on his dinner break at that time and neither saw nor reported this dramatic change in intensity and size of the radar image.

Within the Air Route Traffic Control Center at Fort Worth there was a repeater scope showing the images on the Stephenville radar with a two-minute delay. This was meant to provide data to an ARTC Center weather service meteorologist who had responsibility for creating a center weather advisory (CWA) to inform all controllers of significant changes in weather. With the DFW meteorologist on duty also on his dinner break, no one was there to create the advisory.

At 5:54 p.m., a Feeder East controller broadcast "there's a little bitty thunderstorm sitting right on the final [northeast of the field]; it looks like a little rain shower" (NTSB 1986, p. 47). Delta 191 had switched from his frequency and the flight crew did not hear that transmission. By 6:00 p.m., Delta 191 was being vectored away from the airport, then west, and finally south to be positioned in the stream of landing planes at three-mile (one minute) intervals. The cockpit voice recorder (CVR), when analyzed after the crash, contained comments such as "we're going to get our plane washed" that indicated the crew was aware of nearby rain showers (NTSB 1986, p. 3). Simultaneously, at 6:00 p.m., the radar observer at Stephenville returned to his radar and noted the cell he had seen earlier had reached intensity level 4 (intense rain). One minute later he called the National Weather Service forecast office in Fort Worth suggesting they might want to issue a severe weather statement or a thunderstorm warning based on the rapid growth of this cell.

The forecaster in Fort Worth had no confirmation of severe weather occurring on the ground and did not issue a warning or advisory, though a Level 4 radar return is considered to be an indication of a thunderstorm. During the next four minutes, as Delta 191 flew 12 miles toward the field and descended along the electronic glide slope leading to the runway, there were numerous sightings from pilots and ground-based individuals of rapidly worsening weather and of lightning north of the airport. None of this was reported to the landing aircraft. At 6:03:58 p.m., the captain of Delta 191 called the tower controller and stated "out here in the rain—feels good" (NTSB 1986, p. 3). At the same instant the tower cab supervisor called the TRACON supervisor (in a radar room at the base of the tower) and reported, "We've been busy with these SWAPS [Severe Weather Avoidance Programs] and hadn't paid any attention, but that is heavy, heavy rain off the approach of both runways [18R and 17L]" (NTSB 1986, p. 49).

The TRACON supervisor replied, "Yeah, I can see that" (on his radar) (Belew 1989a, p. 33).

The comments on the flight deck of Delta 191 during the final minutes of its approach to runway 17L at DFW transcribed from the cockpit voice recorder recovered from the wreckage by the National Transportation Safety Board tell a tragic tale. Excerpts from them are listed below (first officer flying) (NTSB 1986, pp. 127–136).

| Time | Source | Individual | Content |
|------|--------|-----------|---------|
| 18:03:03 | Air–ground communications | Approach controller | "Delta one ninety one heavy, reduce your speed to one six zero, please." |
| 18:03:09 | Intra-cockpit | Captain | "One six zero." |
| 18:03:10 | Intra-cockpit | First officer | "All right." |
| 18:03:11 | Intra-cockpit | Captain | "Localizer and glide slope captured." |
| 18:03:16 | Intra-cockpit | First officer | "One six zero is your speed." |
| 18:03:31 | Air–ground communications | Approach controller | "And we're getting some variable winds out there due to a sh-shower on short out there north end of DFW." |
| 18:03:34 | Intra-cockpit | ? | "Stuff is moving in." |
| 18:03:58 | Air–ground communications | Captain | "Tower Delta one ninety one heavy, out here in the rain, feels good." |
| 18:04:01 | Air–ground | Tower | "Delta one ninety one heavy regional tower one seven left cleared to land, wind zero nine zero at five gusts to one five." |
| 18:04:06 | Air–ground | Captain | "Thank you, sir." |
| 18:04:07 18:04:15 | Intra-cockpit | First officer; captain, flight engineer | "Before landing check [verbal exchange]." |
| 18:04:18 | Intra-cockpit | First officer | "Lightning coming out of that one." |
| 18:04:19 | Intra-cockpit | Captain | "What?" |
| 18:04:21 | Intra-cockpit | First officer | "Lightning coming out of that one." |

| Time | Source | Individual | Content |
|------|--------|-----------|---------|
| 18:04:22 | Intra-cockpit | Captain | "Where?" |
| 18:04:23 | Intra-cockpit | First officer | "Right ahead of us." |
| 18:04:30 | Intra-cockpit | Flight engineer | "You get good legs don't ya?"[1] |
| 18:05:04 | Intra-cockpit | ? | "Wash that off a little bit." |
| 18:05:05 | Intra-cockpit | Captain | "A thousand feet." |
| 18:05:08 | Intra-cockpit | Captain | "Seven sixty two in the baro." |
| 18:05:12 | Intra-cockpit | First officer | "Aw right." |
| 18:05:19 | Intra-cockpit | Captain | "Watch your speed." |
| 18:05:20 | Intra-cockpit |  | "(Sound similar to rain begins and continues to impact.)" |
| 18:05:21 | Intra-cockpit | Captain | "You're gonna lose it all of a sudden, there it is." |
| 18:05:26 | Intra-cockpit | Captain | "Push it up, push it way up [referring to engine power levers]." |
| 18:05:27 | Intra-cockpit | Captain | "Way up." |
| 18:05:28 | Intra-cockpit | Captain | "Way up." |
| 18:05:29 | Intra-cockpit | Captain | "Way up." |
| 18:05:29 | Intra-cockpit |  | "(Sound of engines high RPM.)" |
| 18:05:30 | Intra-cockpit | Captain | "That's it." |
| 18:05:36 | Intra-cockpit | Captain | "Hang on to the # #." [expletive] |
| 18:05:39 | Intra-cockpit | First officer | "(What's vee ref?)" |
| 18:05:44 | Intra-cockpit | GPWS[2] | "Whoop whoop pull up." |
| 18:05:45 | Intra-cockpit | Captain | "Toga [take off go around]."[3] |
| 18:05:46 | Intra-cockpit | GPWS | "Whoop whoop pull up." |
| 18:05:47 | Intra-cockpit | ? | "Push it way up." |
| 18:05:48 | Intra-cockpit | GPWS | "Whoop whoop pull up." |
| 18:05:49 | Intra-cockpit | GPWS | "Whoop whoop pull up." |
| 18:05:52 | Intra-cockpit |  | "(Sound of noise similar to landing; sound of takeoff warning horn, the sound continues for 1.6 seconds.)" |
| 18:05:53.5 |  |  | # [expletive]. |

1. "Second Officer's [Flight engineer's] . . . comment was not a casual cockpit comment. His reference to 'good legs" was to legs of the trip, as it was the practice at Delta that the Captain and First Officer alternate flying legs. Second Officer . . . was concerned and was pointing out a deteriorating situation and poor flying conditions" (Belew 1989a, p. 36).

2. Ground proximity warning signal.

3. The captain threw a switch that caused the flight displays to indicate the flight control positions for takeoff or go around maneuvers and set the engine power for those maneuvers, indicating a decision to abandon the approach and attempt to escape the microburst.

In the 37 seconds between the captain's admonition to "watch your speed" given one second before entering heavy rains to the tower controller's cry "Delta go around," Delta 191 was in extremely heavy rain. The words and sounds on the cockpit voice recorder clearly indicate an encounter with a severe microburst that created a headwind followed by a downburst then a strong tailwind too strong for three engines at full power to overcome and maintain flying speed. The detailed analysis of the information on the digital flight data recorder confirmed this and added much detail regarding the events of the final mile and a half of the flight as it crashed.

## MICROBURST

A microburst is a small, vertical downdraft (downburst) of air—usually within a thunderstorm—that results in a horizontal outburst of winds when striking the surface. Horizontal vortices of wind are usually associated with the surface outburst (Figure 5.3). The concept of a microburst was introduced and named by Dr. Tetsuya "Ted" Fujita, while investigating wind damage patterns on the ground after the massive tornado outbreak of 1974. A microburst has outflow low-level winds of less than 2.5 miles in diameter. The winds usually last less than two to five minutes and may reach speeds as high as 168 mph. Their danger to an aircraft during takeoff or landing is in dramatic wind shear.* The typical impact of a microburst on a landing aircraft (Figure 5.3) is as follows:

- As the plane enters (point A), there is a headwind increase from the outflow on the approach side.
- This decreases and there is an abrupt drop in altitude as the plane enters the downburst (point B).
- In a matter of a few seconds, there is a rapid increase in the tailwind component (with a dramatic decrease in indicated airspeed).
- These events in rapid sequences may result in the plane dropping to the ground at high speed while the crew is trying desperately to keep it airborne.

Between 1964 and 1985, 628 people died in 27 microburst-related aircraft accidents in the United States. Other fatal microburst-related crashes oc-

---

* In meteorology, wind shear is the local variation of wind speed in a given direction or wind direction or speed in vertical layers. In aviation, it is the variation over time along the path of an aircraft.

curred at such diverse locations as Okinawa, Australia, Pago Pago, Mexico, Nigeria, Arabia, and India. The highest wind speed recorded in a microburst at an airport occurred on August 1, 1983, at Andrews Air Force Base near Washington, D.C., when a microburst hit ground at the location of the anemometer on the field five minutes after Air Force One landed there with President Reagan on board. Winds of 150 mph from the northwest were quickly followed by winds of 107 mph from the southeast three minutes later.

After his initial identification of microbursts in 1974, Dr. Fujita was persuaded to investigate the fatal crash of Eastern Airlines flight 66 at John F. Kennedy International (JFK) Airport on June 24, 1975, when 112 people died and 12 were injured. He utilized the flight data recordings of 14 aircrafts that attempted to land in sequence on the same runway and determined that there were three microbursts from the complex thunderstorm off the approach end of runway 22L at the airport. His intensive analysis showed the increasing impact of the weather on each successive plane over the 20 minutes preceding the crash as the three successively stronger microbursts descended near the northeast end of the runway.

The importance of microbursts in the vicinity of airports led to two major investigations of microbursts sponsored by the U.S. government under the guidance of the National Center for Atmospheric Research (NCAR), with funding provided by the National Science Foundation (NSF). The first of these projects was begun in 1978 and named Northern Illinois Meteorological Research on Downbursts (NIMROD). A network of 3 Doppler radars and 27 meteorological stations was operated 24 hours a day from May to July 1978. Detailed data were obtained on 50 microbursts that occurred on the 1,200 square mile (approximately 30 miles by 40 miles) test area. All were analyzed in detail. A second field investigation of microbursts was conducted from May to August 1982 near Denver, Colorado, in the Joint Airport Wind Study, using similar equipment; 186 microbursts were observed and analyzed.

These two research projects greatly expanded knowledge of microbursts and led to efforts to detect them at or in the vicinity of airports. A third major investigation of microburst recognition was made in 1984 at the Memphis, Tennessee, airport, utilizing Doppler radar and a dense network of specialized surface weather stations. Dr. Fujita was a principal investigator (working with others) in these three major research efforts. He published a series of technical reports on the findings of these projects. They showed that Doppler radar was able to detect the presence of microbursts and that a network of anemometers measuring wind speed and direction at several locations on or near airports could identify and measure the intensity of low-level wind shear. Though terminal Doppler weather radar (TDWR) was recognized as the choice detector, its high cost prevented its widespread use. After 10 years of intensive research on microbursts, in his 1985 book *The Downburst: Microburst and Macroburst*, Dr. Fujita prophetically predicted the likelihood of future accidents because of a microburst (Fujita 1985). That is precisely what happened to Delta 191 at DFW.

## LOW-LEVEL WIND SHEAR ALERT SYSTEM

After the studies of the low-level wind shear at JFK Airport produced by the interaction between a sea-breeze front and the outflow of inland thunderstorms and involving the crash of Eastern Airlines flight 66 in 1975, the Federal Aviation Administration developed a low-level wind shear alert system (LLWAS) to detect large-scale wind systems. This "phase-1" system was very simple. It compared the wind measured at an airport center field location with the winds at four or five other sensors around the perimeter of the airport. When there was a 15-knot vector difference between any of the sensors, the system would alert the tower controller and indicate the winds at each of the sensors. The controller could then broadcast the winds at each sensor to pilots who then had to estimate their potential headwind and tailwind changes during landing or takeoff. This was the system in operation at DFW on August 2, 1985, which did not alert during the approach of Delta 191. While the system worked for large-scale weather factors, such as frontal passages, the sensors were too far apart to detect significant smaller features, like most microbursts, and it gave many false alarms when the center field wind was variable.

During the 1980s, NCAR conducted extensive research that indicated microburst wind shear was very dangerous to aircraft below 1,000 feet and that a far more complex LLWAS system was needed. In 1983, the FAA asked NCAR to develop an enhanced system that could detect microbursts on or near airports. Between 1983 and 1988, LLWAS versions 2 and 3 were developed. As of 2012, phase-2 LLWAS systems, installed at 100 airports, have only the five sensors in the original system but contain improved computer algorithms that reduce the false alarms. They are not very effective for locating microbursts on or near the field. Phase-3 LLWAS were initially installed at nine U.S. airports and are effective in locating microbursts on or within two miles of the field. Most phase-3 systems have between 12 and 16 sensors. The Denver International Airport has 32 sensors—the highest number of any. The greatest advance in microburst sensing has been made through the installation of over 45 terminal Doppler weather radars by 2012, with many others scheduled for future installation.

The LLWAS system at DFW in 1985 consisted of five sensors—one each at the northeast, southeast, southwest, and northwest corners and a fifth one at the center of the airport. The system would sound an alarm and show the wind speed and direction whenever any of the instruments showed a vector difference of 15 knots from any of the others. At the time of the Delta 191 accident, the microburst winds did not reach any of the LLWAS stations and no alarm was sounded until after the crash. After the accident, the microburst winds moved over the northeast sensor, which did not create an alarm. The winds then moved over the center field sensor, starting them all in alarm at 6:10 p.m. (five minutes after the accident) with winds of 45 knots, increasing to 70 knots at 6:25 p.m., then decreasing to less than 20 knots by 6:32 p.m., as the microburst outflow moved south across the field.

## NTSB

Immediately after the crash, the National Transportation Safety Board "Go Team," made up of specialists on all aspects of aircraft, piloting, airfields, weather, air traffic control (ATC), low-level wind shear detecting systems (air and ground), and every aspect of the accident, were on their way to DFW to study what happened. They gathered information, held hearings, and after a year of intensive study released the National Transportation Safety Board Accident Report AAR-86-05 in August of 1986 (NTSB 1986). The first 55 pages of the report dealt with facts and detailed technical information. The 20 following pages discussed the board's detailed analysis. Their conclusions, finding, and probable cause statements, listed below, were followed by 11 pages of recommendations and 72 pages of detailed technical information.

The NTSB's 23 findings were as follows:

1. Between 1752 and 1800, the Cell "D" radar weather echo positioned off the north end of the DFW Airport intensified from a VIP Level 1 to a VIP Level 4.
2. The absence of the CWSU meteorologist from his station between 1725 and 1810, and the failure of CWSU procedures to require the position to be monitored by a qualified person during his absence precluded detection of the intensification of the weather echo north of DFW Airport.
3. During its final approach to runway 17L, flight 191 flew into a very strong weather echo (VIP Level 4) located north of the field. The weather echo contained a thunderstorm with a heavy rainshower.

4. The thunderstorm produced an outflow containing a microburst. The microburst touched down just north of the DFW Airport. The center of the microbust was 12,000 feet (1.97 nmi) north of the approach end of the runway 17L and about 1,000 feet west of the extended centerline of the runway and the ground track of flight 191.

5. The microburst diameter was 3.4 kilometers. The horizontal windshear across the microburst was at least 73 knots, and the maximum updraft and downdraft were 25 fps [feet per second] (4.8 knots) and 49 fps (29 knots), respectively.

6. There were six distinct reversals of vertical wind components along the southern side of the microburst. The presence of this type of wind flow showed that vortices had formed along the boundary between the descending air and the ambient environment.

7. Flight 191 penetrated the microburst and the vortex flow in the southern side of the microburst.

8. The first officer successfully transited the first part of the microburst encounter by rotating the airplane above a 15° nose-up pitch attitude and by increasing engine thrust to almost takeoff power.

9. About 1805:35, 17 seconds before initial impact, the airplane encountered rapid reversals in the lateral, horizontal, and vertical winds causing the stickshaker to activate [indicating imminent stall]. The first officer exerted a 20- to 25-pound push force on the control column in response to the stickshaker.

10. The flight director was placed in TOGA mode during the initiation of the missed approach 7 seconds before initial touchdown. The flight director's TOGA mode does not command the optimum pitch attitudes required to transit a low-altitude wind shear. However, the Safety Board could not determine whether the first officer was following the pitch commands provided by the flight director's TOGA mode during the final 7 seconds of the flight.

11. The first officer exerted a 20- to 25-pound pull force on the control column in order to avoid ground contact. The stickshaker activated momentarily, and the first officer relaxed the pull force on the control column, which made ground contact inevitable.

12. Delta 191 touched down softly and almost avoided ground contact. [However, it struck a vehicle and skidded into two water tanks, broke apart and burst into flames.]

13. The ATC controller's speed adjustment procedures were not causal to the accident.

14. The 3 nmi separation standard was not maintained between flight 191 and the preceding Learjet. The loss of separation did not contribute to the accident.

15. The Feeder East and Arrival Radar-1 controllers provided flight 191 with all weather information that was available to them.

16. Several flight crews saw lightning in the rain shower just north of the airport; however, they did not report what they saw to the ATC controllers.

17. The LCE controller observed lightning about or shortly after the time flight 191 entered the microburst windfield. Therefore, the failure of the LCE controller to report it to flight 191 was a causal factor.

18. The flight crew and the captain had sufficient information to assess the weather north of the approach end of runway 17L. The lightning observed and reported by the first officer was adequate, combined with the other data known to the flight crew and captain, to determine that there was a thunderstorm between the airplane and the airport.

19. The north side of the cell formation containing the thunderstorm was not masked from flight 191 by any intervening clouds.

20. The captain's decision to continue beneath the thunderstorm did not comply with Delta's weather avoidance procedures; however, the avoidance procedures did not address specifically thunderstorm avoidance in the airport terminal area.

21. After penetrating the first part of the microburst, the engine thrust, which had been increased, was then reduced and at 550 feet AGL the airplane had restabilized momentarily on the glide slope. The captain evidently believed that they had successfully flown through the worst of the microburst wind shear, and the approach was continued.

22. The company had not provided guidance to its flight crews concerning specific limits on the excursions of airplane performance and control parameters during low-altitude wind shear encounters that would dictate the execution of a missed approach.

23. Although the captain did not audibly express his decision to execute a missed approach until he called for the selection of the "TOGA" mode on the flight director 7 seconds before initial impact, maximum engine thrust had been applied before the airplane's rapid departure below the glideslope." (NTSB 1986, pp. 78–80)

These were followed by the NTSB statement of probable cause, which read:

The National Transportation Safety Board determined that the probable causes of the accident were the flight crew's decision to initiate and continue the approach into a cumulonimbus cloud which they observed to contain visible lightning; the lack of specific guidelines, procedures, and training for avoiding and escaping from low-altitude wind shear; and the lack of definitive, real-time wind shear hazard information. This resulted in the aircraft's encounter at low altitude with a microburst-induced, severe wind shear from a rapidly developing thunderstorm located on the final approach course. (NTSB 1986, p. 80)

## LEGAL ACTIONS

Within hours after the accident, individuals, lawyers, insurance companies, and groups anticipating legal suits sprang into action. Weather and air traffic control consultants were contacted to assure their availability for the inevitable trials. There were 180 claims and 150 lawsuits filed against Delta Air Lines in diverse locations. In separate actions, the widows of Captain Connors and Second Officer Nessick sued the U.S. government, alleging negligence by employees of the U.S. National Weather Service and the Federal Aviation Administration, which prevented the flight crews from receiving adequate warnings of the adverse weather they were about to encounter on their final approach to DFW. Delta joined as a plaintiff in these suits against the U.S. government, alleging negligence by employees of the National Weather Service and the Federal Aviation Administration in failing to provide the flight crew warnings of known dangerously adverse weather conditions that had developed north of the airport.

In the microburst-related crash of Pan American Airlines 759 during takeoff from the New Orleans International Airport in Kenner, Louisiana, on June 9, 1982, the U.S. government avoided a liability trial by agreeing that there was a shared responsibility between the parties. In the case of the DFW Delta 191 crash, the government denied they had any liability and claimed any fault was fully because of the negligence of the flight crew. In accord with federal law, all the legal actions filed in the many federal courts throughout the United States were transferred to the single federal court nearest the event. This was the U.S. District Court for the Northern District of Texas, Fort Worth Division. Liability alone would be tried by that court. The initial multidistrict litigation organizational meeting was held in Fort Worth in January 1986. Counsel for all plaintiffs and defendants attended.

The question of damages (which would be tried separately) was addressed by Robert Alpert, the senior vice president and director of claims for U.S. Aviation Underwriters [United States Aircraft Insurance Group (USAIG)]. He announced that anyone who wanted a trial on actual damages could have one immediately and liability would not have to be proven. He assured all the attorneys that no claimant would suffer financially while awaiting trial or settlement of damage claims. Other arrangements were approved by the court whereby the government, if found liable, would reimburse Delta appropriately. Discovery—in which lawyers for each side question witnesses about their potential trial testimony prior to their appearance on the witness stand to "discover" what their testimony will be and prepare cross-examination questions for use during the trial—was held in various legal offices in Dallas and Fort Worth over a period of several months. The court would determine the liability only for the estimated $150 to $200 million in claims as well as the $24,700,000 value of the aircraft.

## TRIAL PREPARATION

On the evening of the crash, potential litigants contacted experts and consultants to alert them of the tragedy and ask them to assemble data and begin analyses of events within their area of expertise. Mr. Robert Alpert, who had previously engaged Climatological Consulting Corporation in the microburst-related crash of Pan American flight 759 in New Orleans to work with Dr. Fujita in the weather analysis related to that accident, alerted me to begin assembling data and to collaborate with Dr. Fujita on a detailed analysis of the weather associated with this crash. Dr. Fujita welcomed opportunities to investigate real-life severe weather events and agreed to investigate the weather circumstances and make his analyses available to all parties in the legal actions by publishing his results. He also welcomed any assistance we could provide. USAIG and Dr. Fujita were aware of our ready access to the archives for the weather data as well as the satellite and radar imagery and requested we provide him with all the data he and his staff required for their detailed analysis.

Walter Bohan, a CCC associate in Park Ridge, Illinois, and a weather satellite expert with access to the Man computer Interactive Data Access System (McIDAS) at the University of Wisconsin–Madison, was able to obtain and analyze the cloud imagery over the several states surrounding DFW and demonstrate their progressive changes in time. I obtained all the surface and upper-air data, including forecast, for the afternoon and evening of August 2, 1985, from the National Climatic Data Center. The weather radar imagery from radars in Oklahoma City, Oklahoma, and Stephenville, Texas (both within range of DFW), were followed at four-minute intervals, but the geographic overlay charts on these radars did not portray the location of DFW. I added that location to the images and provided Dr. Fujita the successive images in the form of time-lapse imagery. Frank McDermott, an air traffic control expert consultant with whom I frequently associated, provided both taped and transcribed copies of the conversations on the cockpit voice recorder and the transmissions to and from the air traffic controller and Delta 191 on the airborne digital flight data records (DFDR) on the aircraft (essential to an analysis of the motion of the aircraft and its control inputs).

With financial support, sponsored by grants from the National Aeronautical and Space Administration (NASA), the National Environmental Satellite Data and Information Service (NESDIS) of NOAA, and NSF to the Wind Research Laboratory at the University of Chicago, Dr. Fujita led a massive research project whose results were published in 1986 in *DFW Microburst on August 2, 1985* (Fujita 1986). Utilizing the radar and satellite data along with ground-based measurements and the detailed aircraft data from the digital flight data recorder, Dr. Fujita determined the characteristics of the parent cloud that spawned the most complicated microburst he had ever analyzed. His analysis showed that the DFW microburst was complex and included a series of at least four separate sequential downbursts. As Delta 191 entered the downburst at A in Figure 5.3, it encountered a headwind, followed by a downflow at point B and then a tailwind at point C, which was followed by a sequence of increasing tailwinds created by a series of near-ground level vortices on the southside of the descending microburst. These tailwinds reached as strong as 70 knots and overcame the plane's ability to fly out of the southern side of the microburst. The book became a principal scientific exhibit—together with separate reports by other researchers—at the ensuing liability trial held in the Federal Court in Fort Worth, Texas.

## TRIALS

There were 2 trials for liability and about 150 trials for damages. A major consolidated trial to determine liability was held in Fort Worth, Texas. With interruptions, it lasted 14 months. In the trial in Fort Worth, 48 witnesses appeared. Their testimony in 1,800 pages of records was supported by 11,460 exhibits. The transcripts and exhibits were used by the plaintiff and defense attorneys to produce briefs and indices that Judge Belew described as "excellent and highly beneficial to the Court" in creating his Memorandum Opinion and Judgment (Belew 1989a, p. 1). He released these on September 1 (four years and one month after the crash of Delta 191).

The separate lawsuit in Fort Lauderdale was filed in May 1989 by the family of Sidney Bernstein, an employee of Delta who was killed in the crash. They sought damages in the amount of wages he would have earned over the next 20 years if he had lived. The trial was held in the Federal Court in Fort Lauderdale before Federal Judge Jose Gonzalez with a six-person federal jury. The attorney for the Bernstein family sought and was granted an early trial for liability and damages. Several of the witnesses who had testified at Fort Worth, including me, were called to testify at Fort Lauderdale.

After getting the weather analyses and related facts into the court record via live expert witnesses, he called Rufus Lewis, the pilot of the Learjet, which was flying 2.5 miles and 59 seconds ahead of Delta 191 during the approach to landing at DFW. Dr. Fujita's analysis of the conditions experienced by the four planes ahead of Delta 191 (which landed) and the two planes behind Delta 191 (which went around) showed that the Learjet had passed under the downward-rushing air of the microburst just before it reached the ground. While on the glideslope his plane had dropped 500 feet (from 1,600 to 1,100 feet above the ground in 10 seconds) and lost 25 knots of airspeed in those same 10 seconds while passing just beneath the downward-rushing air that was hurling toward the ground at 3,500 feet per minute. Pilot Lewis instantly applied full power and put the plane into a steep climb, escaping being driven to the ground. He testified he crossed the runway threshold "hot and high" to land fast far down the runway (NTSB 1986, p. 20). He testified that he was trapped and had no indication of what he was flying into. Attorney Howard Barwich argued that the pilots of Delta 191 were in a similar surprised situation but encountered the microburst 59 seconds later when it had reached the ground and generated the complex series of horizontal vortices that the Delta 191 pilots were unable to overcome.

The jury of the Fort Lauderdale trial deliberated for two days and agreed that Delta 191 was overcome by a very unusual weather event, which surprised the crew. The jury found Delta not liable, and Judge Gonzalez denied the sought-after damages award. If the Bernsteins had not sought liability as well as damages and waited for Judge Belew's verdict in Fort Worth, they would—as did others—have received their sought-after damages award.

On September 1, 1989, Federal Judge David O. Belew released his 72-page Memorandum Opinion and 1-page Judgment (Belew 1989a):

Consistent with the Memorandum Opinion entered in this case this same date, this Court finds that Plaintiffs have failed to prove that the United States of America was guilty of any negligence that proximately caused the air crash of DL 191 on August 2, 1985, and that judgment is entered in favor of the United States of America. Each party shall bear its own costs. IT IS SO ORDERED. (Belew 1989b, p. 1.)

## APPEALS

The plaintiff attorneys, believing the crash had multiple contributing causes, appealed to the Court of Appeals for the Northern District of Texas, seeking that the court grant their request for a rehearing of the case for further appropriate proceedings and consideration of the doctrine of comparative negligence rather than "sole proximate cause" used by the court, which they believed was obsolete and inappropriate. Their appeal was denied by that court. The plaintiffs then appealed to the U.S. Court of Appeals for the Fifth Circuit. They sought a rehearing or a referral of the case to the U.S. Supreme Court on the same basis. The Fifth Circuit Court of Appeals denied both requests, stating that Judge Belew had applied existent choice of law and was a knowledgeable jurist. It is ironic that two federal courts came to opposite conclusions. Judge Gonzalez with the aid of a jury found Delta blameless, while Judge Belew found them at fault. I was pleased to receive a letter of appreciation from Delta's chief attorney, which included the statement "had Judge Belew remembered your testimony, Delta would have won hands down."

## LESSONS LEARNED

Delta 191 was not the last microburst-related commercial airline tragedy. It was followed by a USAir flight crash in Charlotte, North Carolina, in 1994 and an American Airlines crash in Little Rock, Arkansas, in 1999. Delta 191, however, sparked major efforts to advance technology in weather detection and prediction, flight crew training, aircraft design and operations, and even fire and rescue equipment design. Today, DFW has 18 LLWAS sensors located over a vastly expanded network and surrounding the airport. There are two vital terminal Doppler weather radars continually sensing the inner activities of all clouds within miles of the airport. All pilots are trained in simulators to "fly the DFW microburst," more than 18,000 firefighters from 20 countries have been trained to use greatly improved firefighting procedures by DFW's firefighting crew, and communication procedures and equipment have been enhanced throughout the entire aviation industry. The shared objective has become to insure there will never be another Delta 191–type accident.

# CHAPTER SIX

# LAKE COAMO FLOOD

*An Act of God? Puerto Rico, 1985*

## THE EVENT

Torrential rains at the rate of an inch per hour fell for nearly 24 hours in the Coamo River basin in the foothills and coastal plain of semiarid, south-central Puerto Rico during the night of October 6–7, 1985. In the words of the official U.S. Geological Survey Report on the ensuing flood event:

> During October 6–7, 1985, intense rains and severe flooding occurred throughout southern Puerto Rico. As much as 23 inches of rainfall were recorded during a 24-hour period east of Ponce. Rainfall probably exceeded 16 inches throughout most of the Río Coamo basin (see Figure 6.1). Runoff in the Río Coamo basin peaked in the early hours of October 7, resulting in extensive flooding downstream from the town of Coamo toward the coast (see Figure 6.2).
>
> The flood waters in the Río Coamo eroded the eastern abutment of the northern span of the bridge on Highway 52 (Las Americas Expressway). The approach slab to the first span of the bridge collapsed into the floodwaters. (Johnson et al. 1987, p. 2)

Aggravated by the darkness of night, the lack of highway lighting, the torrential rain, and the water on the highway, drivers were unaware of the disappearance of the highway ahead until their cars plunged into the turbulent waters of the Río Coamo and were swept away. Heroic efforts by police and tow truck operators saved several lives. A few travelers escaped from their sinking cars, were swept to shallow water and miraculously survived. Twenty-nine casualties were reported. The rains were also responsible for creating other disasters nearby, including landslides that swept away shanties and buried over 300 people. A collapsed wall in a section of Ponce caused 13 additional deaths. Puerto Rico's governor requested and President Reagan declared Ponce, Coamo, and Santa Isabel (see Figure 6.2), as well as other areas, disaster zones.

## THE GEOGRAPHY

Lying within the tropics, 1,000 miles southeast of Miami, Florida, and 700 miles east of Jamaica, Puerto Rico is a mountain top island in the 1,300-mile, east–west chain of islands making up the Greater Antilles, which separate the North Atlantic Ocean from the Caribbean Sea. They are the above present-day sea level tops of an ancient mountain chain whose bases lie 17,000 to 27,500 feet beneath the sea. Puerto Rico is a roughly rectangular, brick-shaped island about 35 miles north–south and 113 miles east–west. Coastal plains and low, rolling hills, occupying about 30% of the land area, fringe the island. The Cordillera Central—the central mountain chain—rises abruptly on the west coast of the island and extends eastward somewhat south of the east–west midline of the island and terminates in the northeast and southeast interiors. The mountain slopes are quite steep and rise to heights above 4,000 feet. The southern slopes are much steeper than the northern ones, as the mountain tops and ridge lines are only about 10 miles from the south shore and 25 miles from the north shore.

Lake Coamo was created in 1914 by damming the southward-flowing Coamo River to provide water for the irrigation of huge fields of sugar cane and the production of electric power by the Puerto Rico Electric Power Authority (PREPA). When the dam was built, the lake had the capacity to hold 115 million cubic feet of water. Eroded soil was carried by rain water down the steep slopes to the lake and deposited as silt, which reduced the water holding capacity of the lake to 11 million cubic feet by 1964 and 4 million cubic feet by 1970. By 1985, the lake was completely silted and had no water-holding capacity. The spillway originally had a water discharge capacity of 60,000 cubic feet per second (26,929,860 gallons per minute), which was greatly diminished by 1985 because of the vegetation on the silt in the lake impeding water flow at and above the spillway.

Then, 56 years after the lake was created, the four-lane divided highway between San Juan and Ponce was constructed in 1970. In the Coamo area several routes were considered, some inland of the lake and one below (seaward) the lake. This latter route was chosen because of the scenic beauty of the lake and the water flow over the dam spillway as well as ease of construction on the flatter, near-coastal land.

## THE CLIMATE

Mild temperatures, refreshing sea breezes near the coast, frequent brief showers, and plenty of sunshine make the climate of most of Puerto Rico favorable for both tourists and agriculture. The exemption is the arid segment on the south-central part of the island sheltered by the mountains from the persistent trade winds blowing year-round from the northeast. The climate of the island is dominated by the interaction of the prevailing and persistent northeast trade winds and the slopes of the central mountains. These trade winds [blowing from the northeast to southwest at steady speeds of 12–18 knots (14–21 mph)] provided the wind power for Christopher Columbus's armada of 17 ships and 1,500 men who first visited the island on November 19, 1493. These winds normally strike the northern slopes of the mountains and cause frequent showers, while the southern slopes are sheltered with light winds and receive very little rain. The rising air on the northern slopes expands and cools as it moves up the mountain slopes; its moisture is condensed into clouds from which rain falls almost daily as showers and thunderstorms. When the air, depleted of most of its moisture, passes over the ridges and descends toward the southern coast, it compresses, warms, and becomes very dry.

The mean annual rainfall for the period 1971–2000 (the 30-year normal) is shown in Figure 6.3. The dark red areas have the least rain (30 to 50 inches per year), while the green areas have 75 to 105 inches on average. Average annual rain in excess of 140 inches per year is shown in shades of purple. Ponce, on the south shore, west of the Coamo River basin, normally has 55–60 inches of rain, distributed as shown in Figure 6.4 (measured at a station four miles east of the Ponce Post Office), with the greatest monthly amount—7 inches—occurring in October. The Coamo River basin, east of Ponce, is ordinarily the driest area of the island with normal, annual rainfall of less than 30 inches, distributed by months as in Figure 6.5 (measured two miles southwest of the Coamo Post Office). In the mountains north of the Coamo basin, the normal average annual rainfall varies between 70 and nearly 100 inches, while the shore locales on the north side of the island at the same longitude have an average annual rainfall between 50 and 60 inches. The 22 inches of rain in a single day on October 6–7, 1985, over the Coamo River basin was as much as would normally be expected in eight months (two-thirds of an average year) and over three times the amount usually expected in the full month of October.

## PUERTO RICO'S HURRICANES AND TROPICAL STORMS

The Antilles Islands, including Puerto Rico, stretching from Cuba to Barbados, are often in the paths of Cape Verde hurricanes (late-season major hurricanes that form near the Cape Verde Islands off the west coast of Africa). Many of these later strike the U.S. mainland after inflicting great damage on the islands through wind and rain. Twentieth-century hurricanes of this type that severely impacted Puerto Rico, and their statistics are listed in Table 6.1.

**TABLE 6.1.** Twentieth-century Cape Verde hurricanes that severely affected Puerto Rico.

| Date | Name | Wind | Deaths | Damage ($) |
|------|------|------|--------|------------|
| September 13, 1928 | San Felipe II[1] | East, 150 mph | Over 3,000 | Phenomenal |
| August 12, 1956 | Betsy | East, 90 mph | 9 | 25.5 million |
| September 18, 1989 | Hugo | East, 110 mph | 18 | 2 billion |
| September 18, 1996 | Hortense | East, 80 mph | 19 | 153 million |
| September 21, 1998 | Georges | East, 115 mph | 3 | 2 billion+ |

1. San Felipe I occurred on September 13, 1876.

By far the most devastating of those was the September 13, 1928, hurricane, which was named San Felipe II after the saint's day on which it struck. It wreaked havoc on the islands of Guadeloupe, St. Kitts, Montserrat, St. Croix, and Puerto Rico; then it crossed the Bahamas and struck Florida near Palm Beach and moved over Lake Okeechobee, killing over 3,000 people, mostly by drowning. In all these hurricanes, the winds on Puerto Rico came predominantly from the east and the rainfall patterns were controlled by the topography, with the maxima on the north slopes and the mountain ridges. San Felipe II remains today the hurricane with the greatest impact on Puerto Rico.

## THE WEATHER

On October 6 and 7, 1985, a developing, rather weak easterly wave (a westward-moving migratory wavelike disturbance in the trade wind easterlies) approached southern Puerto Rico and became nearly stationary near the western end of the island. The cyclonic, counterclockwise (in the Northern Hemisphere) flow of air disrupted the normal trade wind flow and brought winds from the southeast and south to the normally arid, southern coastal shores of Puerto Rico and the steep south-facing slopes of the inland mountains. It was not a hurricane or a large-scale weather event, but it was a rainmaker. As the weak wave, which had slowly moved westward for several days, turned north over western Puerto Rico, the winds from the south on its eastern side pushed very warm and wet oceanic air up the steep, southern slopes of the backbone of the island, unlike the winds from the east of the more rapidly moving historical hurricanes.

This orographic uplift (see Figure 6.6) created very heavy rains and intense thunderstorms that continued for over 20 hours as they saturated the soils of the steep slopes and rapidly flooded the many tributaries of the Coamo River and the river itself. When the overwhelming flow of water rushing seaward reached the silt-filled bogs and vegetated flats of the former Lake Coamo, they took whatever route they could find, which included the following:

- over the dam spillway and under the Highway 52 bridge (54,000 cubic feet per second; 24,235,200 gallons per minute);
- over the far eastern portion of the levee that "protected" the southern edge of the original lake (18,000 cubic feet per second; 8,078,400 gallons per minute);
- through the Highway 153 underpass under Highway 52 (8,000 cubic feet per second; 3,590,650 gallons per minute); and
- westward along the northern embankment of Highway 52 to the river (10,000 cubic feet per second; 4,488,300 gallon per minute). (Johnson, 1987, p. 1.)

When the water rushing over the levee was forced west by the highway embankment, which was rapidly eroded, the concrete slabs protecting the embankment were washed away. The highway was undermined and a segment of the roadbed nearest the bridge collapsed into the raging waters. After lingering over the western part of Puerto Rico and producing the Coamo River floods, the storm moved northward and intensified into a tropical depression and, briefly, into Tropical Storm Isabel, which retained tropical storm intensity for only two days before approaching the Florida and Georgia coasts on October 9, where it produced no significant damage before moving northeastward and dissipating over the Atlantic Ocean.

## THE LITIGATION

Subsequent legal actions were brought on behalf of the dead, injured, bereaved, or economically affected against institutions and individuals the plaintiff attorneys believed were in some manner accountable for the multiple disasters. As is common practice in such litigation, the plaintiff attorneys, representing the injured families, sued everyone and every agency that they felt was connected to the planning, design, construction, or maintenance of the highway, the lake, the dam, or could be conceived as being in any way responsible for the tragedies on the highway on the night of October 6–7, 1985. Their hope was to collect as much as possible from as many defendants as possible, for the benefit of their clients. Many of the lawsuits were settled out of court. Several were tried in the Federal Court of San Juan and awards totaling $1,699,200 were made in favor of plaintiffs who lost family members when their cars fell into the river through the huge hole in the highway. Other unresolved suits were assigned to the Court of First Instance, Superior Court of Ponce, Commonwealth of Puerto Rico. Many of these were settled out of court in the interval of 1986–1989.

Three years after the event, 19 remaining unresolved cases were tried as consolidated cases in a single trial in the Court of Ponce. Actions were brought against the following:

- the Puerto Rico Highway Authority (PRHA), which owned and was responsible for the maintenance of the highway;
- the Tito Castro Construction Company, which performed some of the original highway construction;
- the Guillermety, Ortiz and Associates Engineering Co. (GTA), which did the original design work on the affected portion of the highway;
- the Commonwealth of Puerto Rico (CPR);
- the Aqueduct and Sewage Authority of Puerto Rico (ASAPR), which had pipes under the highway;
- PREPA, which owned a dam a short distance upstream on the Coamo River above the highway;
- the police of Puerto Rico (PPR);
- various other parties; and

- the many insurance companies that provided coverage to the various defendants.

On November 1, 1990, the trial of the unresolved cases began. There were 41 attorneys—about equally divided between plaintiff and defense—and 45 witnesses (23 for the plaintiff and 22 for the defense) involved in the trial, which continued (with many interruptions) for over three years until March 18, 1994. At times during the trial, individual settlements—totaling over $2 million—were achieved, reducing the number of parties to the suit, but several plaintiffs remained to the end of the trial. Though the plaintiffs, defendants, and attorneys were all from Puerto Rico, there were many technical experts in engineering, weather, hydrology, construction, human factors, and related disciplines engaged by the plaintiffs and defendants to assist the court in its search for the truth in our adversarial system of justice. These came from as far away as Georgia, Kansas, Massachusetts, and North Carolina.

A few years before the Coamo flood, I worked with highway design and construction engineer Richard Schwab, employed by Howard, Needles, Tammen, and Bergendoff (HNTB). We testified during litigation related to a major flood in Fargo, North Dakota, and respected each other's technical knowledge and ability to communicate scientific facts to a judge and jury. The defense attorneys, Alex Gonzalez for the PRHA (highway) and Juan Vilella for PREPA (electric power), engaged HNTB to provide a team of experts in engineering of dams, bridges, and highways as well as the hydrology and hydraulics of highway drainage systems. They needed the assistance of an expert on climate and the statistical likelihood of rainfall amounts in specific Puerto Rico locations. They turned to Climatological Consulting Corporation for forensic meteorological and hydrological consultants. I agreed to serve and urged the addition of CCC associate Dr. Vance A. Myers, recently retired as the director of the NOAA Division of Hydrometeorology and an expert on floods, to the team. The attorneys agreed and we went to work learning everything possible about Puerto Rican rainfall and floods as well as the details of the Coamo flood of October 6–7, 1985.

To assure full awareness of all the pertinent details of the rainfall affecting Puerto Rico in general and the Coamo River basin in particular, Dr. Myers and I reviewed all the recorded data for early October 1985 for Puerto Rico together with the climatic history of the island. These included the following:

- the weather maps for October 5–8 covering a very large area centered on Puerto Rico and analyzed at 6-hour intervals;

- the surface weather observations taken at hourly intervals at all weather stations on Puerto Rico for the same dates;
- daily rainfall and temperature data for all measuring stations over Puerto Rico for those same dates;
- copies of the original autographic paper recordings of automated rainfall measuring stations for those four days on the island;
- meteorological satellite imagery for those dates covering a large area around Puerto Rico;
- reproductions of the images of the scope of the U.S. National Weather Service Weather Surveillance Radar at San Juan, Puerto Rico, for October 6–7;
- all official reports pertaining to the weather events, the flood, and their impacts;
- summaries of historical climatological data for Puerto Rico;
- histories of past tropical storms and hurricanes impacting Puerto Rico; and
- the transcripts of the depositions taken of potential witnesses to be called at trial (as they became available).

The weather maps, observations, data, satellite and radar imagery, and climate records were obtained from the NCDC at Asheville, North Carolina. The tropical storm and hurricane histories came from numerous textbooks, such as I. R. Tannehill's *Hurricanes: Their Nature and History, Particularly Those of the West Indies and the Southern Coasts of the United States* (Tannehill 1944).

Together we assembled all the historical rainfall data from all stations in Puerto Rico, analyzed all prior high rainfall events, and performed the appropriate statistical analyses. We read the testimony of all witnesses to the event in prior litigation; visited each of the observing sites; explored the topography of the Coamo River basin; investigated the locations of landslides, structural damage, and so on; and examined rainfall recording equipment, some of which had malfunctioned during the October 6–7, 1985, torrential rains. As is often the case in an extreme weather event, some normally reported data were missing—either because of equipment malfunction or human absence at critical times. Other data that would have been of great value did not exist because there were areas where neither observers nor equipment were located. The central mountains of the island blocked the beam of the National Weather Service radar at San Juan, making it of little use in calculating the rainfall south of the mountains. Our detailed analysis indicated the rainfall on the 58-square mile Coamo River basin had a statistical expectancy of less than two one-hundredths of 1% of occurring in any year, a more than "500-year storm" (one such storm—on long-term average—every five centuries or twice in a millennium). This was far beyond the "50-year storm" considered in the design of the bridge and the highway approaches.

The opposing attorneys have the right to depose the expert in advance of trial and discover the scope and potential impact on their case, providing them an opportunity to prepare their cross-examination questions, often with the assistance of their expert in that field of expertise. Dr. Myers and I were extensively examined in depositions in San Juan well in advance of the Ponce trial. The process of discovery and depositions prior to trial streamlines the court proceedings and greatly reduces the likelihood of "Perry Mason"–like surprises in the courtroom. After extended depositions of both of us, the attorneys for both sides agreed only one of us would be required to appear at trial as we were in full agreement on our prospective testimony. I was asked to testify.

The call to appear at trial came while my wife and I were vacationing on Sanibel Island, Florida. In some courts, once seated in the witness chair, the expert witness may not confer further with any parties to the litigation until finished with their testimony and excused by the court. Such was the court's ruling in the trial in Ponce. I was forbidden to talk with any of the attorneys or other witnesses until all questioning was completed and I was excused by the judge. This restriction during the several days "in the witness chair" made me an isolated person, and my daily calls to Sanibel began with "this is the prisoner of Ponce."

Since the courtroom proceedings were conducted in Spanish, a translator was assigned to the non-Spanish speaking weather witness. Many of the defendants attempted to convince the court that the rainfall that created the floods was so unprecedented, unforeseeable, and overwhelming it could be construed as an act of God (an unprecedented, overwhelming, and unforeseeable "force majeure" unaffected by the intervention of man—a truly unnatural event), absolving individuals and institutions of culpability. But, as is customary in our adversarial system of justice, the plaintiffs led off the proceedings in court. They attempted to introduce evidence that the defendants were negligent in planning, designing, constructing, and maintaining the lake, highway, and their environs, and they should have anticipated the occurrence of the rainfall that created the flood.

Ordinarily the plaintiffs would produce fact and expert witnesses to testify. The fact witness would testify as to what they saw, heard, felt, or otherwise sensed but could offer no opinions. The expert witness would follow and could—in response to questions posed by the attorneys—offer opinions in their area of expertise based on the facts in testimony and their knowledge, training, and experience. After direct examination, in which the expert offered opinions, they would be subjected to cross-examination by the opposing attorneys whose questions would be designed to limit the impact of—or cast doubt on—the validity of those opinions. The cross-examination is supposed to be limited to the topics in the direct examination. Further testimony is elicited from the witnesses in redirect examination, recross-examination, and so on, until all possible clarifications of the details of earlier responses are achieved. The expert witness testimony is meant to further assist the court in its search for the truth. The expert is restricted to responses to questions posed in direct and cross-examination and is not permitted to volunteer information beyond the scope of the questions put before him or her.

It is customary for attorneys and experts to confer extensively in advance of the trial and to create direct examination questions to be asked of the expert at trial as well as cross-examination questions to be asked of their counterpart engaged by the opposing attorneys. While the attorneys are advocates for their clients, the expert witnesses are impartial specialists whose sole role is to provide facts and opinions in their field of expertise that may be useful to the court. While the plaintiff attorneys produced a host of fact and expert witnesses in engineering and related fields who offered fact and opinion testimony regarding their perceived inadequacy and errors in the design, construction, and maintenance of the lake and highway, they did not produce any witnesses regarding the climate of Puerto Rico or the details and/or foreseeability of the weather events of October 6–7, 1985. After being queried for my credentials, education, and experience, including prior acceptance as an expert witness, I extensively went through voir dire by the plaintiff attorneys and was accepted by the court as qualified to testify as a consulting meteorologist.

My direct examination testimony in answer to questions by Defense Attorneys Gonzalez and Vilella took two and a half days of testimony, illustrated with over 50 maps, charts, diagrams, tables, and graphics on vugraph transparencies projected on a screen and explained in detail to show reconstruction of the weather event and the likelihood of such rainfall occurring at the Coamo River drainage basin. The highway and bridges were designed to withstand a "50-year storm," that is, a rainstorm whose statistical expectancy in any year is 2% (two such storms could be expected—on long-term average—each century). When constructing highways, it is customary to design culverts and bridges to safely carry water as "usually expected," according to the likelihood of experiencing various amounts of precipitation. Small two-lane highways in rural areas with little expected traffic may have drainage culverts designed to carry water flows expected to be exceeded once in 25 years. Larger two-lane roads with more traffic often are designed to have drainage systems whose capacity would be exceeded, on average, only once in 50 years. More recent four-lane major highways are ordinarily designed with drainage systems to safely carry away water from rainfall expected to occur only once in 100 years. The "recurrence interval" of various rainfall amounts can be calculated by the analysis of years of past rainfall events.

On February 25, 1994, the plaintiff attorneys began their weather cross-examination, which became extensive and was interrupted after two and a half days by a two-week court recess. It ended on March 15, 21 days after it began, with the two-week interruption. The plaintiff's attorneys did not engage a similar expert witness for weather, and they attempted to place into evidence weather testimony favorable to them by means of questions directed to the defense expert outside of the scope of the direct testimony. Though this is not generally permitted by courts, in this instance the presiding judge allowed it, and it greatly prolonged the aggressive cross-examination. They produced textbooks with rain maps showing the extremely heavy rainfall over the high mountains of the Cordillera Central of Puerto Rico related to Hurricane San Felipe II of 1928—the heaviest rain in Puerto Rico's history—and asked that those amounts (far greater than any of October 6–7, 1985, but well removed from the Coamo River basin) be read into the record.

Without being directly asked about this, I was unable to comment that these rains were well removed from the Coamo River basin and this was scientific nonsense. They argued to the court incorrectly that such rainfall should have been anticipated in the design of Highway 52 at the Coamo River bridge. Though the defense attorneys failed to object to these improper tactics, it appeared the judge was aware and did not consider their argument valid. The court, after presiding over three years of often conflicting testimony, ruled that an "act of God defense" required proof of an unprecedented, overwhelming, and unforeseeable force *unaffected by the intervention of man*. The design and building of highways, dams, bridges, pipes, and other structures (especially the highway embankment) the court deemed to be "intervention of man" and ruled in favor of the plaintiffs, whose financial compensation would be individually determined in subsequent damage trials for each of the remaining plaintiffs.

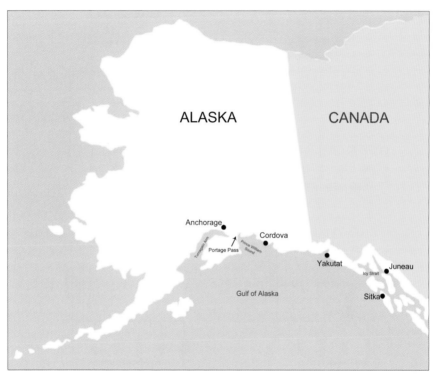

**FIGURE 1.1.** Flight route: Anchorage to Juneau, Alaska.

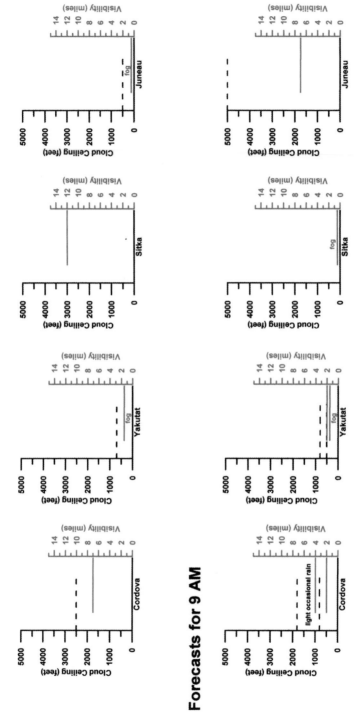

**FIGURE 1.2.** En route weather observations and forecasts.

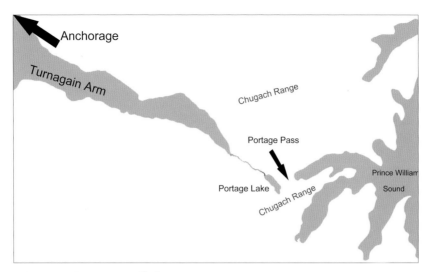

**FIGURE 1.3.** Portage Pass, Alaska.

**FIGURE 3.1.** Map of Florida, Tampa Bay, and the Skyway Bridge.

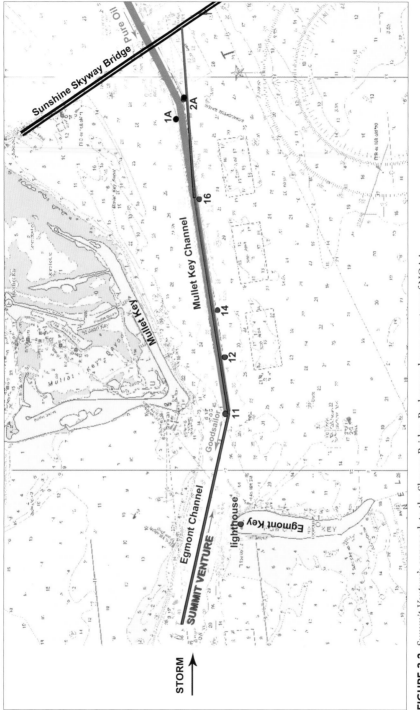

**FIGURE 3.2.** *Summit Venture's* approach to the Skyway Bridge. Background map courtesy of NOAA.

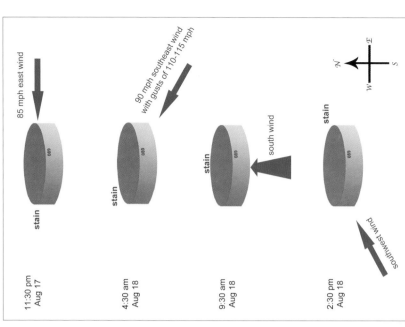

**FIGURE 4.2.** Tank 089 failure timeline.

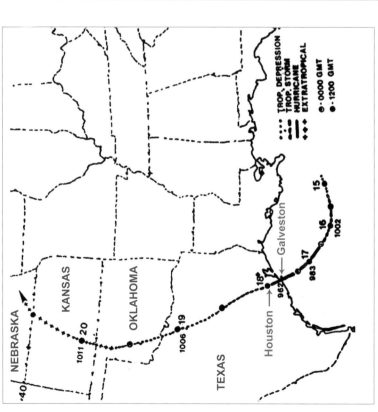

**FIGURE 4.1.** Hurricane Alicia storm track August 15–21, 1983. Courtesy of NOAA's National Centers for Environmental Prediction (see www.wpc.ncep.noaa. gov/tropical/rain/alicia1983trk.gif), modified by the author.

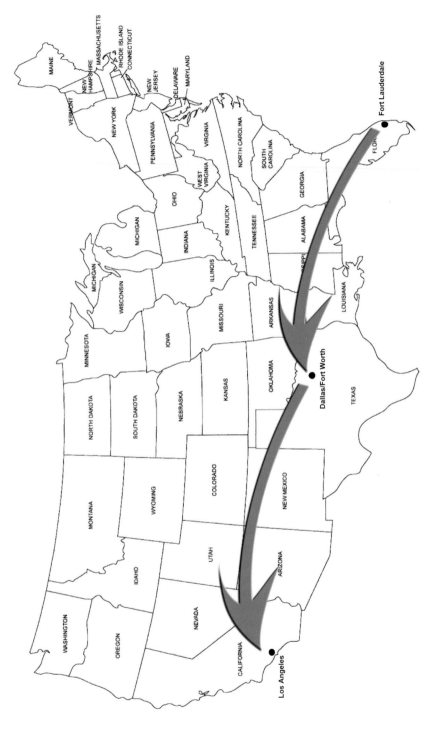

**FIGURE 5.1.** Map of Delta 191 planned route. Background map courtesy of Bruce Jones Design, Inc., and FreeUSandWorldMaps.com, modified by author.

**FIGURE 5.2.** Map of Dallas–Fort Worth area.

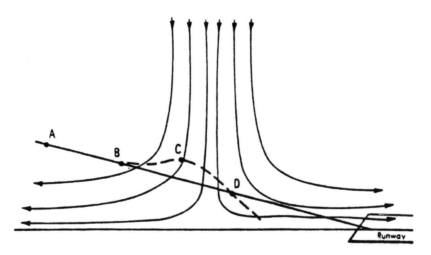

**FIGURE 5.3.** Typical impact of a microburst on a landing aircraft. Points A–D indicate the position of the aircraft. Courtesy of the National Transportation Safety Board NTSB/AAR-86/05.

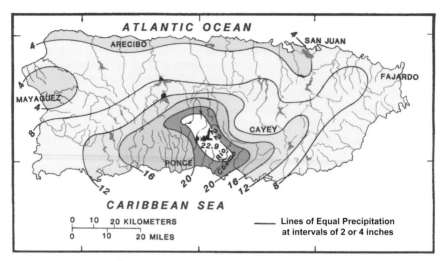

**FIGURE 6.1.** Cumulative precipitation over Puerto Rico during October 6–7, 1985. Courtesy of the Department of the Interior, USGS, Open-File Report 87-123 (modified by the author).

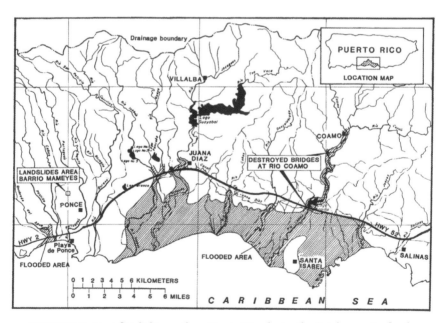

**FIGURE 6.2.** Areas flooded in southern Puerto Rico during the October 7, 1985, flood. Courtesy of the Department of the Interior, USGS, Open-File Report 87-123.

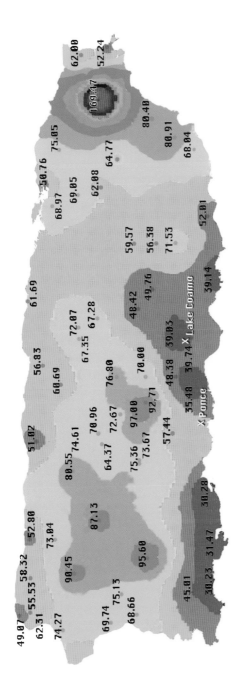

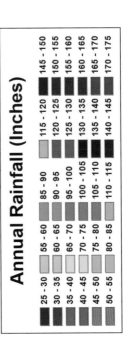

# Annual Rainfall (Inches)

| | |
|---|---|
| 25 - 30 | 85 - 90 | 145 - 150 |
| 30 - 35 | 90 - 95 | 150 - 155 |
| 35 - 40 | 95 - 100 | 155 - 160 |
| 40 - 45 | 100 - 105 | 160 - 165 |
| 45 - 50 | 105 - 110 | 165 - 170 |
| 50 - 55 | 110 - 115 | 170 - 175 |
| 55 - 60 | 115 - 120 | |
| 60 - 65 | 120 - 125 | |
| 65 - 70 | 125 - 130 | |
| 70 - 75 | 130 - 135 | |
| 75 - 80 | 135 - 140 | |
| 80 - 85 | 140 - 145 | |

**FIGURE 6.3.** Puerto Rico mean annual precipitation 1971–2000. Courtesy of the NWS, modified by the author.

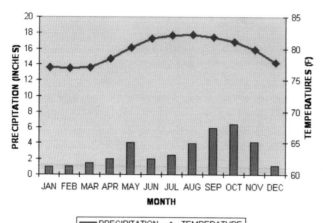

**FIGURE 6.4.** Average annual rainfall by months at Ponce 4E. Courtesy of the NWS.

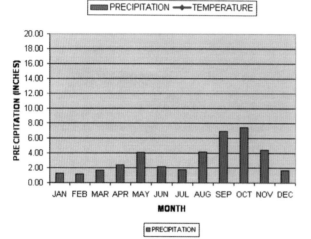

**FIGURE 6.5.** Average annual rainfall by months at Coamo 2 Southwest. Courtesy of the NWS.

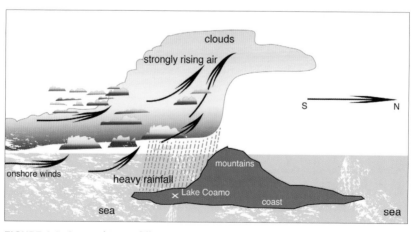

**FIGURE 6.6.** Orographic rainfall.

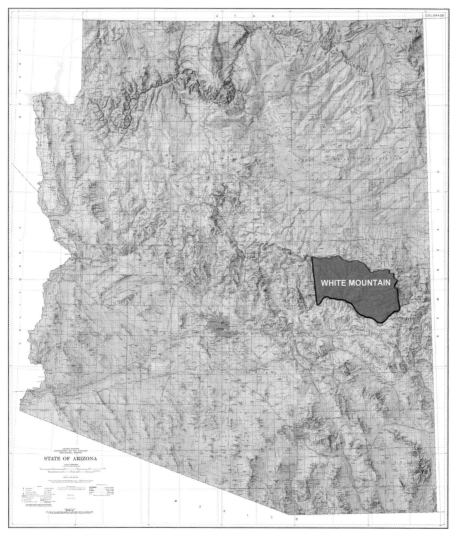

**FIGURE 7.1.** Location of White Mountain Apache Reservation in Arizona. Courtesy of the United States Department of the Interior Geological Survey, modified by the author.

**FIGURE 8.1.** Planned route of flight from Salt Lake City, Utah, to Roswell, New Mexico.

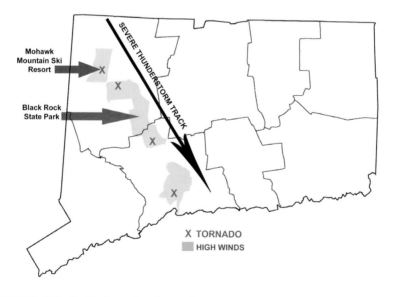

**FIGURE 9.1.** Track of high winds and tornadoes across Connecticut on July 10, 1989.

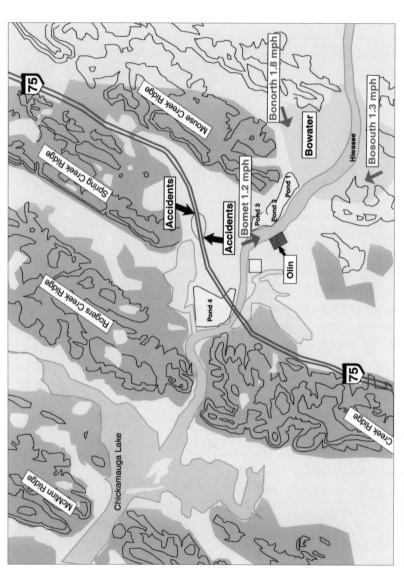

**FIGURE 10.1.** Site topography.

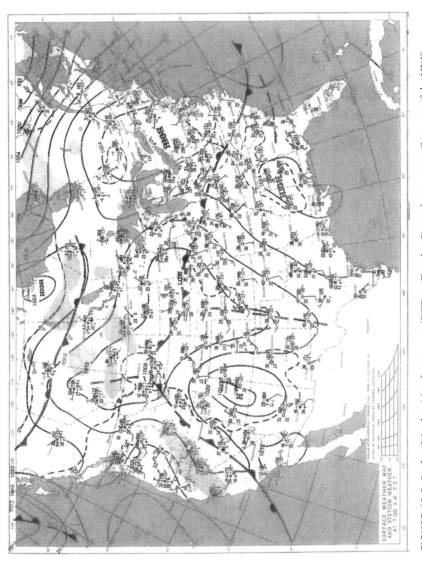

**FIGURE 10.2.** Surface Weather Map for 7:00 am (EST) on Tuesday, December 11, 1990. Courtesy of the NWS.

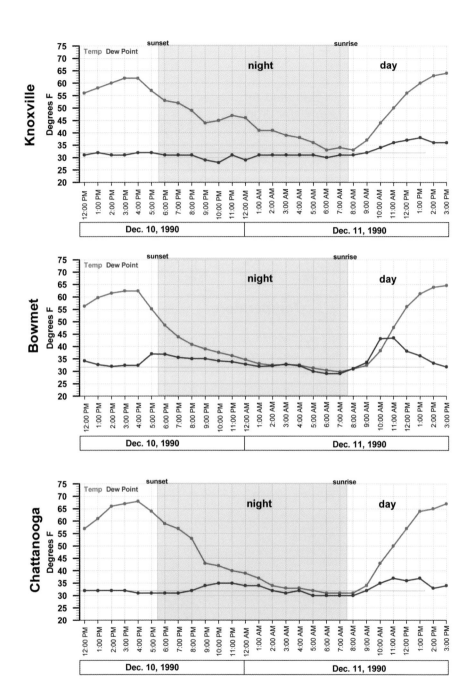

**FIGURE 10.3.** Hour by hour temperature and dewpoint.

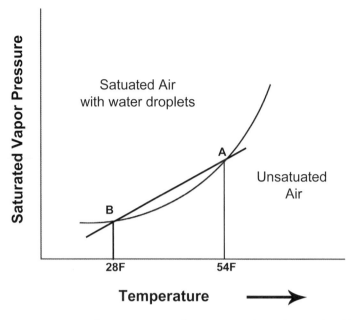

**FIGURE 10.4.** Saturated vapor pressure and temperature relationship. Produced by George McVehil, this graph was presented in a report that became part of the deposition testimony.

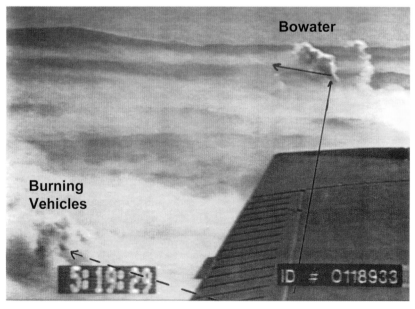

**FIGURE 10.5.** Frame 29 of aerial video on the morning of December 11, 1990. Source: Bowater gave permission to use during deposition.

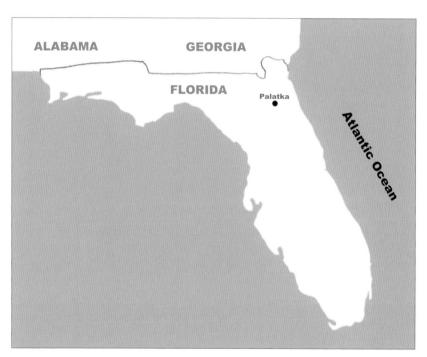

**FIGURE 11.1.** Site location.

**FIGURE 12.1.** Flight route from Covington, Georgia, to Gainesville, Florida.

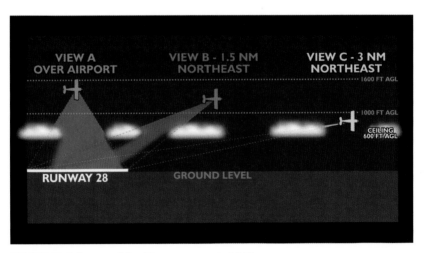

**FIGURE 12.2.** Impact of cloud layer on ground visibility.

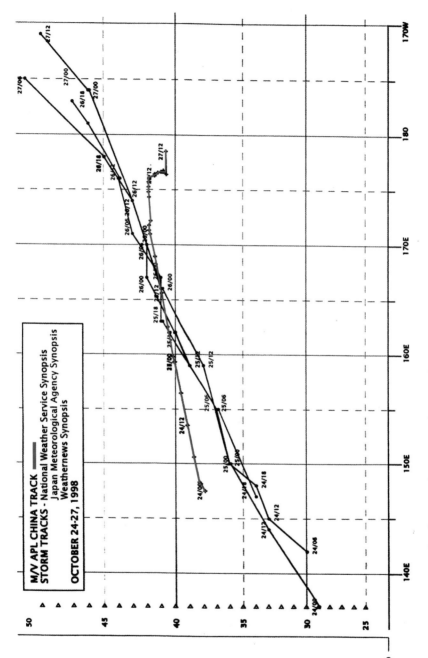

**FIGURE 13.1.** Tracks of storm and M/V *APL China*, October 24–27, 1998.

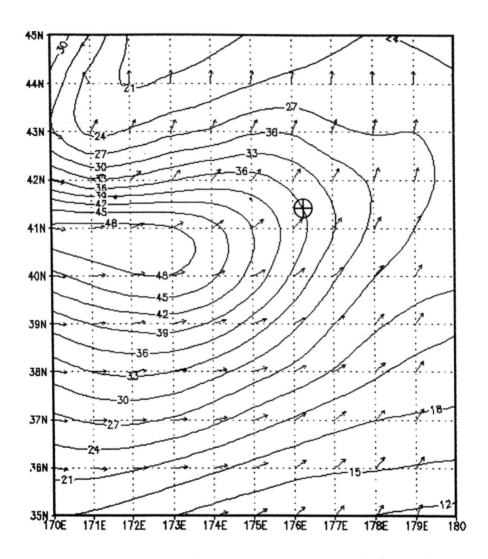

**FIGURE 13.2.** Segment of significant wave height and primary wave direction chart for 1200 UTC on October 26, 1998. Prepared by U.S. Navy Fleet Numerical Weather Central, Monterey, California; produced by Fleet Numerical Meteorology and Oceanography Detachment (FLENUMMETOC DET) Asheville.

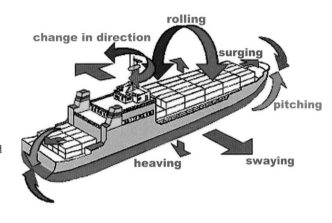

**FIGURE 13.3.**
Ship movement at sea. Courtesy of "Container Handbook: Cargo loss prevention information from German marine insurers," modified by the author.

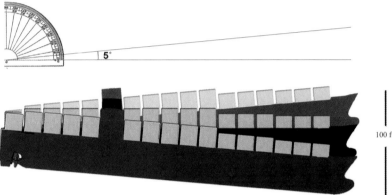

**FIGURE 13.4.** 5° pitch.

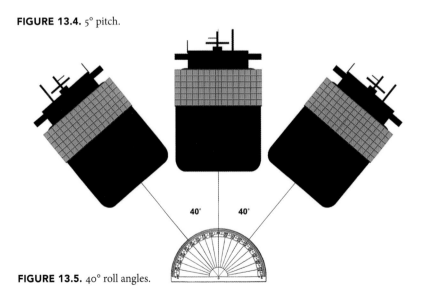

**FIGURE 13.5.** 40° roll angles.

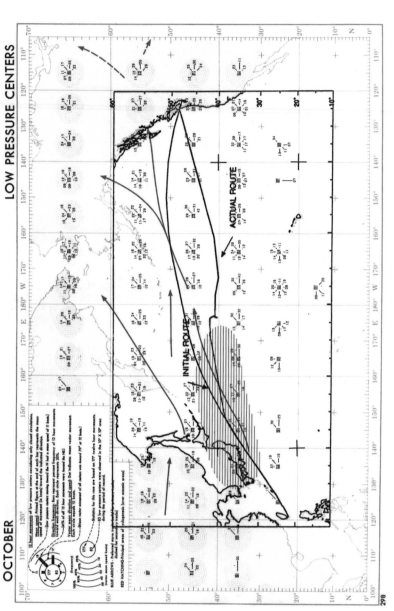

**FIGURE 13.6.** Area of cyclogenesis with planned and actual track of ship. Courtesy of the U.S. Department of Commerce, Weather Bureau, U.S. Department of Navy, Navy Hydrographic Office (1959). *Climatological and Oceanographic Atlas for Mariners* (Vol. 2). Washington, DC, USA.

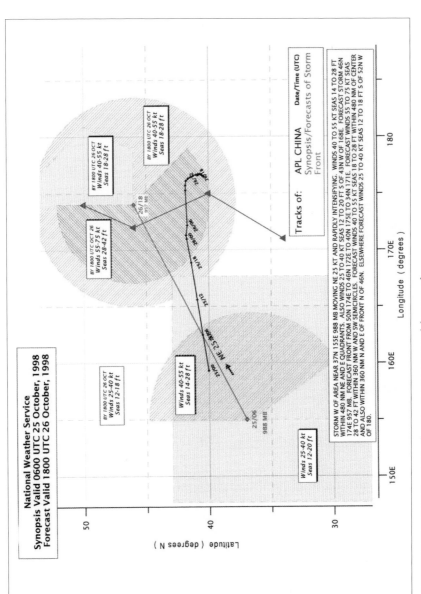

**FIGURE 13.7.** U.S. National Weather Service synopsis and forecast of storm.

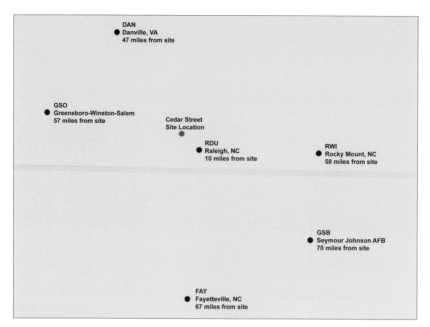

**FIGURE 14.1.** Weather station location in relation to the site.

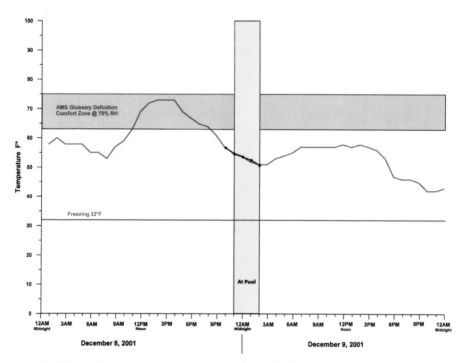

**FIGURE 14.2.** Temperatures at Raleigh, North Carolina (in red), and interpolated temperatures at the site (in black) on December 8 and 9, 2001.

# CHAPTER SEVEN

# WHITE MOUNTAIN APACHE TRIBE V. USA

*Mismanagement, Arizona, 1870-1985*

## EARLY SETTLEMENTS

It is widely believed among scientific groups that the earliest humans to settle in what is present-day Arizona migrated by foot across the land bridge between Asia and North America, which today is as much as 400 feet beneath the surface of the waters of the Bering Sea. At that ancient time, 20,000 to 10,000 years ago, the polar regions of Earth were covered by ice as much as two miles thick and the sea surfaces were hundreds of feet below their present level. The early migrants wandered slowly, over many generations, eastward and southward, gradually occupying much of the land areas of North, Central, and South America. Large groups reached what is now Arizona, New Mexico, Colorado, and Utah and became known as ancient Pueblo people.

At first mostly nomadic hunter-gatherers, they gradually transitioned to farmers and lived in communities. With time the climate changed and the lower lands in the southwest of North America increasingly became drought stricken, causing the villagers to seek higher and more productive farmlands. Deserts slowly spread over the lands below 4,000 or 5,000 feet above sea level, and the Pueblo people farmed at and above 6,000 feet. Scientists reconstructing their history have described some as cliff dwellers, remnants of whose abodes exist today. Many were referred to as Anasazi, and although there was never an Anasazi tribe, later scientists used that Navajo title when writing of the people who lived in the "Four Corners" area before AD 1300.

Today's Navajo, Zuni, and Apache Indians are descendants of these ancient Pueblo people whose history has been extensively studied in recent years. The members of the White Mountain Apache Tribe in northeastern Arizona are among those descendants. Many thousands of years later, beginning after the discovery of America by European explorers and the creation of colonies on the east coast, rapid immigration and population growth occurred. It was not long before the newcomers began spreading westward into lands already occupied by the Indians, wrongly named such by Christopher Columbus who thought he had sailed to India when he arrived in the Caribbean Islands.

## THE RESERVATION

In the 1860s, the U.S. Army Cavalry and many warring Indian tribes battled over possession of land in the Arizona Territory. In 1869, Major John Green, the local U.S. cavalry troop commander, made peace with the Apache leaders in the east-central portion of the territory and reached agreement for the creation of a U.S. military post and an Indian Reservation near the confluence of the east and north forks of the White River. Construction began on Camp Ord on the White Mountain River in 1870. Renamed Camp Mogollon, then Camp Thomas, and finally Camp Apache, it became Fort Apache in 1879. The Indians lived on the adjoining lands that were formally established 21 years later, as the Fort Apache Indian Reservation, by presidential executive order in November 1892.

The reservation is now known as the White Mountain Apache Indian Reservation. It lies 194 miles northeast of Phoenix, Arizona, and contains 1,664,000 acres or approximately 2,600 square miles. The northern and southern boundaries are between 30 and 40 miles apart, while the eastern and western boundaries are separated from each other by about 75 miles (see Figure 7.1). The northern boundary of the reservation is along the Mogollon Rim, a very prominent feature of the Arizona topography. The Mogollon Rim—with an average land elevation of 7,000 feet—runs west-northwest to east-southeast nearly 200 miles across central Arizona. It separates the arid and hot southern portion of Arizona from the far more temperate climate of the Mogollon Plateau north of the rim. Water falling on the plateau drains westward to the Colorado River, while water south of the rim drains to the Salt River, which provides the water for present-day Phoenix and the extensively irrigated land of southwestern Arizona.

The southern boundary of the reservation is along the Salt River. All the creeks and rivers on the reservation drain southward to the Salt River, which flows through Phoenix and eventually to the Colorado River. The east and west reservation boundaries run north–south between the Mogollon Rim and the Salt River at the eastern and western edges of the lands occupied by the tribe at the time the reservation was created. It occupies an area of sloping topography and complex climatic patterns. The land slopes from the 2,600-foot elevation near the desert bottom of the Salt River Canyon on the southwest edge of the reservation to the 11,400-foot pine-forested elevation of Mount Baldy near its northeast corner. Less than 11 inches of rain per year falls on the arid southwest of the reservation, while over 40 inches of precipitation—much of it as winter snow—falls on the mountain slopes in the northeast.

Whiteriver, the major population center of the reservation, at an elevation of 5,280 feet, is the capital of the White Mountain Apache Tribe. It enjoys a moderate four-season climate. Over 12,000 members of the White Mountain Apache Indians live on the reservation. Another 3,000 members live and work in other parts of the United States and the world.

## THE BUREAU OF INDIAN AFFAIRS

The Bureau of Indian Affairs (BIA) is one of the oldest bureaus of the federal government. The Continental Congress in 1775 created a Committee on Indian Affairs headed by Benjamin Franklin. When the U.S. Constitution was drafted, Article I, Section 8 empowered Congress "to regulate commerce with foreign Nations, and among the several States, and with the Indian Tribes." The Bureau of Indian Affairs was established by Secretary of War John C. Calhoun on March 11, 1824. Its function was to "oversee and carry out the Federal government's trade and treaty relations with the tribes." In 1849, the BIA was transferred to the Department of the Interior (DOI), where it operated under a variety of sequential names as the Indian Office, the Indian Bureau, the Indian Department, and Indian Services. It was not until September 17, 1947, that it again became officially the Bureau of Indian Affairs. Its functions have varied with time and its current role is as "a partner with tribes to help them achieve their goals of self-determination." The BIA's current mission is "to enhance the quality of life, to promote economic opportunity, and to carry out the responsibility to protect and improve the trust assets of American Indians, Indian tribes and Alaska natives" (www.bia.gov/WhoWeAre/BIA).

## THE SUIT

On October 18, 1950, the White Mountain Apache Tribe filed a suit against the U.S. government seeking $1.25 billion in compensation for alleged long-term and continuing mismanagement of the tribe's water, rangeland, and timber resources. The suit was to be tried in the Indian Claims Commission. There it became Document 22, where it languished without action. Nine years later, on May 25, 1959, the commission severed Document 22 into several cases. The alleged mismanagement became 22-H. The tribe promptly filed an amended petition still seeking the $1.25 billion claim for reimbursement for mismanagement of the tribe's water, rangeland, and timber resources by the BIA. Once again, there was no action. Seventeen years later, on December 16, 1976, this 22-H portion was transferred to the U.S. Court of Claims. In 1985, after another 10 years (27 years after the second amended petition and 36 years after the initial suit was filed) the claims were tried in the U.S. Claims Court in Washington, D.C., before Federal Judge Christine Cook Nettesheim.

## PLAINTIFF'S PERCEPTION OF THE SUIT

The plaintiff's attorneys believed—and attempted to show—that the Secretary of the Interior and his delegates in the BIA deliberately administered the tribe's water, rangeland, and timber resources—indeed the entire reservation—as a source of water for the downstream Salt River Federal Reclamation Project and not for the benefit of the tribe.

### Water

The Salt River Federal Reclamation Project was administered by the Department of the Interior, as was the BIA. The project developed a series of dams and water control structures designed to benefit developers of the lands now surrounding Phoenix and others served by the waters of the Salt and Verde Rivers. Congress passed the Reclamation Act in 1962 to reclaim wastelands through irrigation. Under the act, the Secretary of the Interior administered the irrigation project by contracting with local associations of landowners. The Salt River Valley Water Users' Association (SRVWUA) was incorporated in 1903 and has been the entity contracting with DOI since 1907. They have been "voracious for every drop of water that comes off the reservation" (Nettesheim 1987, p. 17). It became politically powerful.

There may have been as many as 3,000 to 5,000 acres (according to various surveys made through the years) of potentially irrigable land (using pumps and mobile sprinklers) on the reservation. These were never approved nor funded. The Indians—from time to time—did irrigate as much as 300 acres, using gravity flow ditches and side channels that were periodically made unusable by damage caused by heavy rains in frequent summertime severe thunderstorms. The plaintiff attorneys claimed the BIA intentionally discouraged the tribe from attempting to improve their irrigation facilities or expand their irrigated lands so that more water would be available to the SRVWUA.

## Rangeland

The plaintiff's attorneys further claimed the BIA attempted to deliberately overgraze the pasture lands to increase the flow of water off the reservation into the Salt River for the benefit of the downstream users and to the detriment of the tribe. From the clump grass at the edge of the Sonora Desert in the southwest to the midreservation, lush grass lands to the pine forests on Mount Baldy in the northeast, virtually all of the 1,664,000 acres of the reservation were potentially suitable for the grazing of cattle, horses, or other livestock. Well before any substantial livestock grazing began on the reservation, the American Southwest, and Arizona in particular, became the locus of a rapidly developing range livestock industry. Overstocked herds in western Texas swelled into the Arizona Territory as early as 1876. By 1885, Governor F. A. Tritle likened Arizona to "one vast grazing field" and a "stock-raisers paradise." His annual report stated that "the entire grazing area is nearly if not fully occupied where water can be obtained" (Nettesheim 1987, p. 48).

In the next few years, drought combined with continued overstocking took a toll on the condition of the ranges. The grass supply diminished to the extent that the cattlemen were unable to mature prime beef on the pasture lands and were forced to sell their starving steers as "feeders" at greatly reduced prices to buyers from Montana and California who fattened them for market. The 1893 Arizona governor's report noted such action was one option, while another was to try to use the relatively ungrazed lands of the White Mountain Apache Indian reservation.

The history of non-Indian cattle grazing on the reservation began with incidents of trespass in 1891. The Commissioner of Indian Affairs attempted unsuccessfully to have the cattle removed; then agreed to leave the cattle on the reservation and tried to collect a grazing tax from the owners. By 1901, at least 1,000 non-Indian cattle were grazing on the reservation. Little erosion from overgrazing was known to exist at that time. A permit system was created in 1901 to create income for the tribe but was difficult to manage as cattle owners were reluctant to report the number of cattle on the reservation and the U.S. Army troops had great difficulty in policing the range or determining the livestock numbers. At that time little was known either about rangeland carrying capacity (number of 1,000 pound animals who could graze without damaging the rangelands) nor of the actual number of cattle and horses on the reservation. The plaintiff's attorneys argued the BIA was aware the reservation was being overgrazed, allowing erosion to channel water off the reservation for the benefit of the powerful downstream water users, and that the BIA endorsed that to the detriment of the tribe. Between 1900 and 1910 there was increasing evidence of range destruction throughout Arizona because of excess cattle denuding and trampling the rangelands, resulting in erosion. Most of the cattle were not owned by the Indians.

## Timber

The plaintiff's third set of mismanagement claims focused on their belief that the BIA deliberately overharvested timber to denude the extensive forested areas of the reservation, causing increased runoff to provide more water for the downstream water users. The northern and eastern regions of the reservation, from 6,000 feet above sea level to the timber line between 10,000 and 11,000 feet above sea level, are forested with over 1 million acres of open to dense woodland, some being mixed forest and pasture; 685,000 acres of this are suitable for "commercial forest." The forest area today is suitable for "multiple land use," including recreation, water, wildlife habitat, grazing, and timber. The plaintiff attorneys claimed the government owed the tribe for:

- timber harvested and used by the U.S. Army between 1870 and 1917;
- timber cut by the BIA and processed through government-owned and private sawmills;
- damages caused by the government's improper handling of forest fires; and
- the alleged damages caused by the government-prescribed burning of natural debris on the forest floor.

In addition, the plaintiff's attorneys claimed the government required the tribe to overcut to the extent that they forced later reduction in tribal income by reducing the forest's future harvest. They claimed the government's entire forest management scheme and practice throughout the history of the reservation were motivated by the desire to increase the streamflow from the reservation for the benefit of downriver water users.

Timber harvesting on the reservation began in the 1870s for the lumber needs of Fort Apache. It was not until 1918 that timber sales to outside users began. The first formal Forest Management Plan was written in 1944 by the BIA. It was frequently amended or revised as timber harvesting methods changed rapidly from local sawmills to a fixed sawmill location with rail tracks laid through the forest and logs carried by train to "truck logging," where logging roads are built and trucks carry logs to the sawmill. (A sawmill built on the reservation burned in 1947 and was replaced in 1961 by a mill at Whiteriver. Today, that mill is the largest employer of tribal members.) Meanwhile, the BIA created periodic Forest Management Plans based upon the observed growth characteristics in test areas scattered throughout the forests and the periodically changing, widely accepted silviculture methods, which the plaintiffs claimed were not representative of the tribal forests. The plaintiffs continuously alleged all the BIA's forest management plans were designed to overcut the forests for the deliberate purpose of increasing runoff to increase the stream flow for the benefit of downriver water users.

## DEFENSE'S VIEW OF THE LITIGATION

The U.S. Department of Justice was charged with defending the BIA against these claims. The defense attorneys believed the plaintiff's allegations were without merit and sought to show that the BIA had consistently acted in the best interest of the tribe.

## THE TRIAL

Judge Nettesheim ordered each set of attorneys to engage experts to prepare reports on the issues before the court, to exchange the reports, to conduct interrogatories of all the witnesses expected to appear at trial, conduct depositions, and produce all potential trial witnesses. Since the alleged damages—if real—had occurred over a time span of many decades, it was not possible to retain percipient witnesses (those who observed the events). The plaintiff and defense attorneys had to rely on experts who could reconstruct the histories of past events impacting the status of conditions on the reservation and their changes over the past century and determine which were natural and which—if any—were the result of human decisions. Both sets of attorneys quickly assembled teams of experts in irrigation, crop management, hydrology, forestry, meteorology and climatology, grazing land management, geomorphology, and related topics.

All experts were to review all documents of the BIA related to the management of the reservation from its creation to 1985. They were required to make personal visits to the reservation and observe its present condition and any residual indications of prior conditions. Each expert was also to prepare a written report on their findings, review the reports of opposing experts and prepare comments regarding any differences of opinions, submit to depositions conducted by the opposing attorneys, and prepare to testify in direct examination and cross-examination at trial before Judge Nettesheim in the Court of Claims in Washington, D.C. Prior to trial, the judge would review all documents regarding the management of the reservation, all expert reports and their depositions, and make her own personal site inspection of the reservation.

## THE ROLE OF METEOROLOGY AND CLIMATOLOGY

Since weather has both long- and short-term impacts on many of the factors involved in all the operations of the reservation, the attorneys for the defense (i.e., government attorneys) sought potential expert witnesses to testify regarding weather and climate conditions throughout the history of the reservation. In December 1985, Attorney James Upton from the U.S. Department of Justice engaged the services of Climatological Consulting Corporation to review the weather climate and hydrologic history on the reservation and agree to testify factually on behalf of the government. A letter of agreement was signed on December 16, 1985, spelling out the scope of work Climatological Consulting Corporation was to perform under contract to the U.S. Department of Justice. Several associates of CCC began research immediately on the weather and climate of Arizona, reviewing voluminous historical documents on the year to year management of and conditions of the rangelands and forests of the reservation, the stream flow of the rivers originating on the reservation, and multiple relating topics. The work was divided between four members of the firm who worked as a team. All were board-certified meteorologists.

Dr. Myers and I were chosen to be testifying experts at trial, while Henry Harrison and Dr. Werner Baum were to be (together with us) contributing authors and editors of our formal report. All four of us had multiple degrees in meteorology, had fulfilled government and private careers in meteorology, had served in the military (in meteorological assignments during World War II), had been elected as fellows of the American Meteorological Society, had published peer-reviewed articles in professional journals or textbooks on various aspects of meteorology, and were recipients of government and/ or professional society awards. Henry Harrison had been the meteorologist with Admiral Byrd on the first Antarctic Expedition and had been the director of meteorology for United Air Lines. Dr. Werner Baum had a long academic and professional career, having been the president of three colleges and the deputy director of the Environmental Science Services Administration (now NOAA), president of the American Meteorological Society; and chairman of various commissions or boards of the National Science Foundation, the National Research Council, and the University Corporation for Atmospheric Research. Dr. Vance Myers had been an observer forecaster, consultant, and researcher for the U.S. Weather Bureau; consultant on hydrology and floods for the World Meteorological Organization; and a specialist in hydrology, floods, and extreme rainfall. I had retired from 32 years of federal service in meteorology: 6.5 years in the U.S Navy and 25.5 years in the U.S. Weather Bureau/NOAA, of which the last 13 were as the director of the National Climatic Data Center in Asheville, North Carolina, after which I formed Climatological Consulting Corporation in 1976.

We reviewed all of the archived and/or published records of weather and climate pertinent especially to the southwestern United States and to Arizona and the surrounding areas, such as that provided by historical records of climate variations revealed by studies of such items as tree-ring variability in both living and older trees (dendroclimatology). Several meetings were held with other experts whose research was in any way related to past weather events. During our research and analysis of climate and weather variability throughout the history of the reservation, we maintained a close working relationship and collaboration with Dr. William Graf, a geomorphologist from the Arizona State University at Tempe. His analysis of sediment particle sizes in the banks of streams on the reservations needed to be closely correlated with the chronology of major historical rainfall events affecting the reservation. In mid-April 1986, Attorney James Upton transferred from the Department of Justice to the Environmental Protection Agency and turned over his duties on this case to Attorneys Bernard Sisson and Harry Kelso, with whom all government-engaged experts continued their work.

Then, in mid-July 1986, after six months of intensive research, the team of eight potentially testifying defense experts engaged by the government met at Show Low, Arizona, on the reservation for an intensive site visit, including a low-level (above ground) overview by aircraft and helicopter, visits to dozens of ground locations at specific places on the reservation, and interviews with BIA officials. Departing from Show Low Airport with all the defense experts in a twin-tail, eight-passenger, World War II, twin-engine Beech aircraft (fully loaded), we viewed the entire reservation. With a temperature of 104°F in Phoenix and nearly 80° at the 4,000-foot elevation of the Show Low Airport, the high "pressure altitude" for the flight and the weight of the passengers plus pilots plus fuel load approached the operational limits of the plane until some fuel was burned off—especially as we neared the upper slopes of Mount Baldy—over 11,000 feet above sea level. The overview of the reservation from less than 1,000 feet above the ground made it possible to see the streams, irrigation areas, grasslands, forests, and major features of the entire reservation.

On the next day of our visit, we were divided into two groups, each of which boarded helicopters that flew into and landed at over a dozen specific locations on the reservation where we could individually examine rangeland agricultural test plots—fenced off to prevent disturbing vegetative growth and allow crews to record progress. Many of these had weather recording instruments to provide relevant environmental data. We visited grazing cattle herds and examined stream bank erosion, irrigating ditches, timbering operations, and all major activities carried out by the Indians. The opportunity to do this in the company of the experts in other scientific disciplines and exchange ideas "on the spot" was of great value to us all. In the evening we met as a group with reservation staff where we had opportunities to discuss what we had seen and heard and obtain answers to our many questions. The opportunity to discuss our new knowledge among our group of divergent but interrelated sciences proved beneficial.

During August and September 1996, drafts of reports were exchanged (via the attorneys) between all experts, critiques of the reports were created and exchanged, depositions were taken of all experts and exchanged for review and comment, and conferences were held for discussions between attorneys and their consulting experts. The report prepared by the Certified Consulting Meteorologists at CCC, titled "A Report on Climatic Factors; Climatic History; and Climatic Variability Affecting the White Mountain Apache Indian Reservation," (Haggard 1986) was submitted to the U.S. Department of Justice in March 1986. It compared the climate of the White Mountain Apache Reservation with climates of the world, the Rocky Mountain region of the United States, and the complex climate pattern of Arizona. The spatial and temporal variability of several climatic parameters was discussed in detail and the 135-year history of weather observations and the impacts of varying observational networks on climatic analyses were reviewed. Significant historical weather events, such as periodic droughts, heavy localized rainfall in summer thunderstorms, winter mountain snow accumulations affecting stream and river water flow rates, and water availability for pasture lands and irrigation, were analyzed in detail. Several key elements of the weather and climate history of the reservation included the following:

- Winter and summer were characteristically wet, while spring and fall were dry.
- July, August, and September are the wettest months with frequent, and often intense, thunderstorms.
- Though February is often the wettest of the winter months, August rainfall rates are usually nearly double the February rates.
- Major drought prevailed from 1890 to 1904.
- The years 1905 and 1941 were the wettest years of record.
- There was a general (though often interrupted) decline of rainfall from 1920 to 1959, followed by increased values.
- Long-term river runoff was historically quite variable.
- The period from 1905 to 1920 was the wettest in all known history, while the latest prior comparable period was from 1600 to 1620.
- Major droughts occurred in prior centuries and especially 1565–1600 and 1870–1890. Both were followed by wet periods lasting two decades.

The report was accompanied by a color video showing the intensity and geographic distribution of droughts in Arizona from 1895 to 1983.

Thousands of pages of deposition testimony were reviewed by attorneys and experts and provided to the court. Each expert provided their attorney suggested questions and answers for both their examination in direct examination at trial and of the opposition witness at trial. The trial was held in Washington, D.C., during October 1986 with 10 expert witnesses testifying before Judge Christine Cook Nettesheim. At trial Dr. Vance A. Myers discussed the hydrological aspects of precipitation, runoff, and streamflow, and I testified about climate-specific weather events and drought episodes. While the bulk of our testimony occurred in a courtroom, my testimony on drought intensity and location was conducted "in chambers" with the aid of a large screen television utilizing the color video prepared to show the chronological sequence of droughts and wet periods. The judge and I sat facing the television screen with the plaintiff and defense attorneys grouped behind us, interjecting objections and questions from time to time. In her findings, the judge described the video testimony as eloquent.

## LEGAL OPINION

On February 6, 1987, the judge filed her 103-page opinion in which she reviewed the facts and testimony presented at trial and concluded the following:

> It is found and concluded that (1) plaintiff failed to prove by a preponderance of evidence that defendant mismanaged the reservation's water resources, and (2) that plaintiff is entitled to $5,424,429 for the Government's breaches of fiduciary duties in managing plaintiff's rangeland and forest resources. It is further found and concluded that plaintiff did not adduce evidence of any continuing wrong in respect to its water, rangeland or forest resources. (Nettesheim 1987, p. 102).

The judge did, however, award to the plaintiffs $16,698,818.32 additional damages for the mismanagement of the tribe's forests, rangeland and funds, bringing the total award to $22,123,247.32, substantially less than the $1.25 billion sought.

## TRIBAL ACTION

The Tribal Council adopted a resolution to use these funds "to return tribal ecosystems to a condition that better reflects their condition prior to suffering mismanagement and to fund the education of Tribal members in disciplines related to natural resource management" (White Mountain Apache Land Restoration Code 1989, p. 3). They established a "Land Restoration Fund" and invested the damages award into it with the objective of preserving the capital and using the income to finance land improvements projects and educational scholarships and internships in improved land management procedures. The White Mountain Apache Tribe has utilized the income from the invested fund—together with funds it has received from several grants—to undertake a series of land restoration projects. The Tribal Council believes the tribe lands are rich in natural history and bounty and have opportunities for substantial development. The council mandated that the funds be used to support the reconstruction of their tribal ecosystems to conditions representative of their past traditions, including:

- restoration of depleted Apache trout populations;
- watershed restoration, incorporating projects that promote traditional cultural practices;
- restoration of grazing lands with carefully designed irrigation programs;
- forest restoration with special attention to protection of sacred sites, burial grounds, and areas of significant archeological importance;
- wildlife restoration, considered "rewilding," such as the Mexican Wolf Recovery Program, which had traditional significance to wolf rituals practiced by historic warriors;
- cultural reconstruction, including restoration of the use of historic Apache place names; and
- "passive restoration," such as opposing placing telescopes or other modern technologic items on historic cultural sites.

Portions of the funds were especially reserved for the education of young tribal members on the cultural history of the tribe. Tribal members have been extensively involved in the selection of restoration projects to assure the consideration of early tribal culture and assure the compatibility of all restoration projects with historic customs and beliefs.

# CHAPTER EIGHT

# WEBB, RIDING, AND CHARLESWORTH V. USA

*Whiteout at Roswell, New Mexico, 1988*

## THE EVENT

On Friday, February 5, 1988, at approximately 3:00 p.m. (mountain standard time), a Piper Archer aircraft crashed in an open field covered in snow about seven miles northwest of the Roswell, New Mexico, airport. The pilot, Alan Charlesworth, and his two passengers, Lynn Webb and Richard Riding, were killed.

## HISTORY OF THE FLIGHT

The three occupants had chartered the plane for a business trip from Salt Lake City, Utah, to Roswell, New Mexico. Alan Charlesworth, who had very recently acquired his private pilot license and had accumulated only 122 hours of flight time, was licensed to fly only under visual flight rules (cloud ceilings above 1,000 feet and visibility 3 miles or greater). On February 4, 1988, Charlesworth phoned the FAA Flight Service Station serving Salt Lake City at 1:04 p.m. and requested a general outlook for the weather on February 5 for a private plane flight from Salt Lake City, Utah, to Roswell, New Mexico (see Figure 8.1), leaving at about 5:30 a.m. on February 5. The FAA briefer advised that VFR weather would not exist in the Salt Lake City area at 5:30 a.m. and that the weather along the route would be generally good and improving; he did not mention a major snowstorm occurring at Roswell, and he advised the pilot to call the next morning and check on the then current and forecast weather. The pilot called the Flight Service Station a second time at 4:33 a.m. on February 5 and requested a "standard briefing" for the weather from Salt Lake City to Roswell. He spoke to the same briefer.

The FAA computer system that contained the weather details was "down" and had been so all night, so the briefer had no new details. He told the pilot he could only provide general comments: that fog and haze would prevent an early morning VFR departure from Salt Lake City; that when that cleared by midmorning, the weather looked flyable as far as Farmington, New Mexico; and there was some inclement weather near Roswell that he should check on before getting nearer to his time of flight. He provided no details. In fact, there was a major snowstorm in progress at Roswell that would make VFR flight impossible, but the severity of the weather was not mentioned.

The pilot placed a third call to the Flight Service Station at 7:17 a.m. on February 5. He reached a different briefer and indicated he was in the plane ready to go but fogged in. The briefer said the fog would probably clear around 11:00 a.m. and the route would have good weather after that. The briefer was not aware that the pilot had never received a forecast with detailed information about the snowstorm in progress at Roswell. In fact, the Roswell airport was closed. The sky was obscured and the visibility only one half mile, with 10 inches of snow on the ground. Pilot Charlesworth made a fourth call to the Flight Service Station at 8:41 a.m. by radio from the plane, which had become airborne at 8:30 a.m. The pilot requested to change the anticipated flight time on his flight plan from five to seven hours but did not discuss the weather, apparently believing it would be good once he left the Salt Lake City area.

The plane landed at Coronado airport about nine miles north of Albuquerque, New Mexico, in good VFR weather at about 1:00 p.m. to refuel. While on the ground, the pilot called Larry Wright, the plane's owner, in Salt Lake City to obtain permission to have the cabin speaker, which malfunctioned, repaired. Mr. Wright said okay, and while on the phone told Alan Charlesworth that he had heard there was an active weather system over southern New Mexico and advised him to check the Roswell weather before continuing the flight. The pilot agreed to do so, but there is no record of his having contacted anyone about the New Mexico weather while on the ground or in the air after leaving Coronado. The plane departed Coronado about 1:30 p.m. and was in contact with the Albuquerque Air Route Traffic Control Center, which vectored the flight toward Roswell for the next 23 minutes.

The FAA tape of the communications was recycled, and there is no record of the pilot/controller discussion during that part of the flight. At 2:31 p.m., the pilot called the Roswell Flight Service Station and obtained the then current weather data, which had been observed at 1:48 p.m. as ceiling variable at 1,000 feet overcast, visibility of 5 miles in fog, temperature of 33°F, dewpoint of 28°F, wind from 100° at 5 knots, altimeter setting of 30.43 inches of mercury, and ceiling variable of 500 to 1,400 feet. (A more accurate description would have been ceiling 500 feet, broken clouds 1,400 feet overcast, which would have indicated the existing IFR weather conditions at Roswell, for which the pilot was not qualified.) The next observation, six minutes after that conversation indicated the cloud ceiling was 600 feet (IFR). The controller did not contact the pilot with that essential updated information, leaving the pilot to believe the weather at the field was VFR. The briefer failed to warn the pilot of the snow on the ground, the rural nature of the approach, and the danger of a "whiteout" condition. Whiteout is defined as the following:

An atmospheric optical phenomenon [often in polar regions] in which the observer appears to be engulfed in a uniformly white glow. Neither shadows, horizon, nor clouds are discernable; sense of depth and orientation is lost; only very dark, nearby objects [if present] can be seen. Whiteout occurs over an unbroken snow cover and beneath a uniformly overcast sky, when, with the aid of the snow blink effect [a bright, white glare on the underside of clouds, produced by the reflection of light from a snowcovered surface], the light from the sky is about equal to that from the snow surface. Blowing snow may be an additional cause. This phenomenon is experienced from the air as well as on the ground. (AMS 2012b.)

At 2:48 p.m. Alan Charlesworth called the Roswell tower air traffic controller and reported he was 24 miles northwest of the field, flying at 9,500 feet (above sea level but only about 2,800 feet above the ground level of 5,800 feet) inbound for a VFR landing. The tower controller gave the pilot the new weather report and asked the pilot's intentions. At 2:48:48, the pilot replied, "Landing at Roswell, full stop." The controller responded, "The field is IFR" (Greene 1994, p. 23). There was no response, so at 2:49:29 the controller repeated, "The field is IFR." The pilot then said, "The field is IFR . . . OK we are VFR, can you direct us to another airport?" The controller apparently did not hear (or comprehend) that transmission and after several seconds delay asked, "What are you requesting?" Pilot Charlesworth changed his request and responded, "We're requesting a special VFR landing," (Greene

1994, p. 23) though he must have been aware there was IFR weather (which he was not licensed to fly in) in the 24 miles between his location and the airport. The controller surprisingly granted the request. The controller had not previously, as required, provided a Notice to Airmen (NOTAM) to the pilot, but then did so, advising that Runway 21—7,100 feet long—had been plowed 100 feet wide for 5,000 feet and that the snowbanks on either side of the plowed area were 28 inches tall.

These several conversations between the controller and the pilot occurred between 2:49 and 2:59 p.m., as the plane continued to approach the field. During that time individuals on the ground had seen the plane flying toward the airport about 300 feet above the ground (6,000 feet above sea level), "straight and steady" with between three and four miles of horizontal visibility. They lost sight of it when it was about eight miles from the airport. The plane crashed, headed in the opposite direction very shortly after these people lost sight of it. Evidently, the pilot, seeing the snow and the worsening weather ahead, had decided to reverse his course and abandon the approach to the airport. He attempted to make a 180° turn and had crashed during the turn, killing all three men. There were no witnesses to the crash, which occurred in open pastureland, providing no visual ground references on the unbroken snow surface atop a 17-inch layer of snow on the ground.

The representatives of the estates of the three men killed in the crash brought suit against the U.S. government, the Federal Aviation Administration, the Piper Aircraft, Inc., the company that leased the plane, and various companies that provided flight instruments on the plane. They were seeking damages for financial losses in the amounts of the loss of future potential lifetime earnings. The representatives of the two passengers initially brought suits against the pilot as well. The initial suits against all the original defendants other than the United States were settled or withdrawn between 1988 and 1992. The families and estates of the passengers Lynn Webb and Richard Riding were represented by Joel M. Allred of Salt Lake City, Utah, and the family of the estate of Alan Charlesworth were represented by Attorney Moses Lebovitz of Los Angeles, California. The U.S. government, as the defendant, was represented by Attorney Luke B. Marsh of the Justice Department and Mark B. Baylen of the Federal Aviation Administration.

The plaintiff attorneys, Lebovitz and Allred, agreed to work together as a team working out of Joel Allred's office in Salt Lake City, as the trial would be held in Salt Lake City. Both sets of attorneys set up teams of expert witnesses to appear as witnesses at trial as well as fact witnesses such as the FAA Flight Service briefers who had provided weather data and forecasts to the pilot, eyewitnesses who saw the plane in flight approaching Roswell, and others involved in any way with the event. The team of experts for both plaintiff and defendant included experts in air traffic control, piloting, weather, finance, and other specialties, nearly all of whom were deposed by the opposing attorneys during the months of trial preparations.

## PERSONAL INVOLVEMENT

It was not until November 22, 1991, 34 months after the accident, that I became aware of the accident and its legal circumstances. On that date, an air traffic control expert, Andy Hayes, who I had frequently worked with on aviation litigation, called me on the phone. He told me of the crash and the developing legal aspects. He was engaged as a consultant and potential witness in a pending trial against the United States by Los Angeles, California, Attorney Moses Lebovitz, who was representing the heirs of Pilot Alan Charlesworth. Mr. Lebovitz had asked Andy if he could recommend a forensic meteorologist who might testify about the weather matters in the pending trial. Andy Hayes told him about me and offered to see if I was available.

Shortly thereafter I received a call from Moses Lebovitz asking me to work on the case and asking me a number of technical questions about the procedures of taking weather observations, especially ceiling and visibility. I promptly acquired all the weather data that would have been available at the time of a flight from Salt Lake City, Utah, to Roswell, New Mexico, on February 5, 1988, plus the National Transportation Safety Board accident investigation report, which contained copies of the transcripts of all contacts between the pilot and FAA personnel at the Flight Service Stations and air traffic control facilities. I reviewed all the pertinent information and developed several opinions, which I shared with Attorney Lebovitz. I soon learned that he and Joel Allred, representing the estates of the passengers, were working together and that I would be working for them jointly. After reconstructing the weather and reviewing the transcripts of all communications, I reported my several opinions to the two attorneys. By this time many of the original suits had been resolved or consolidated and there would be a single trial versus the United States to be held in Federal Court in Salt Lake City.

I was invited to meet the two attorneys at Joel Allred's office in Salt Lake City to discuss my work product and review my opinions. I did so in early 1992, and I was informed they would soon make me available to be deposed by the attorneys for the United States and the Federal Aviation Administration. My deposition was set for June 18, 1992, in Mr. Allred's office in Salt

Lake City and took place there on that date. Luke Marsh, attorney in the U.S. Justice Department, was the deposer. He was accompanied by Mark Baylen, attorney for the Federal Aviation Administration, and the weather consultant hired by the government. Attorneys Lebovitz and Allred and I, plus two shorthand reporters, met in Joel Allred's office for the deposition.

Attorneys Marsh and Lebovitz were the protagonists, while their weather consultant assisted the U.S. attorney off the record with suggestions for questions. In answer to questions by Mr. Marsh, I testified as to my qualifications, the data and information I had acquired, and the opinions I had reached. These included details about the weather that was recorded; the official forecasts made by the government and by a private forecasting company, Weather Services International (WSI); and the communications between the pilot and a series of five Flight Service Station weather briefers at the Flight Service Station at Cedar City, Utah, serving Salt Lake City, and the air traffic controller at Roswell. None of the weather briefers had provided essential information about a record-breaking snowstorm underway at Roswell, or had provided misinformation indicating good weather there, leading the pilot to undertake or continue the flight into the deadly weather that led to the crash. Further questions concerned the misunderstanding of the weather observations and rapidly varying weather conditions in Roswell that were misinterpreted.

It was my belief that the Flight Service Station briefer provided the pilot inadequate information about the weather during the first call on February 4. He indicated the weather would be flyable once away from the Salt Lake City area but failed to mention the heavy snowstorm at Roswell, New Mexico, the flight's destination. The same briefer was again negligent at the time of the early morning request for an update on February 5. He misinterpreted the graphic weather chart from Weather Service International and believed it indicated good weather at Roswell, rather than the major snowstorm underway there, and he wrongly repeated his earlier statement that the weather over the later part of the flight would be flyable. In fact, the field at Roswell was closed at that time with half a mile of visibility and 10 inches of snow on the ground. The next briefer also failed to advise the pilot of the New Mexico snowstorm during the phone call at 7:19 a.m. on February 5. Once again, when the pilot opened his flight plan by radio from the plane after takeoff, the FAA personnel failed to advise him of the nonflyable weather ongoing at Roswell.

I testified that I believed that the recurrent misinformation about the weather at the flight's destination gave the pilot the belief that once away from the early morning fog and clouds in the Salt Lake City departure area, the remainder of the flight would be in good weather, though a heavy snowstorm was in fact continuously occurring at Roswell, with over 24 hours of non-VFR weather and the field there being closed much of the time. Further, when the pilot neared Roswell and contacted the air traffic controller at the Roswell airport, he was given inadequate, though available, information on the status of the field by the failure of the traffic controller to advise the pilot of the several NOTAMs as required by government regulations. All of these incorrect and or incomplete communications misled the pilot to fly into inclement weather, which ultimately led to the crash. I later assisted the two plaintiff attorneys in preparing questions to ask the government's weather expert in his deposition scheduled at a later date.

## THE TRIAL

The trial was held without a jury from March 1 to 19, 1993 (five years after the crash), in the Federal District Court for Salt Lake City, Utah, before Federal District Judge Thomas Greene. The judge heard live testimony from the FAA employees, as well as from all other witnesses, and reviewed their deposition testimony. The lawyers offered many exhibits prepared by their witnesses and experts. Final arguments were made by the attorneys on March 19, the last day of trial. The attorneys were allowed to file post-trial memoranda and make further submissions to the court. The judge reviewed all the evidence and the submissions of the attorneys and then made his findings of facts, conclusions of law, and judgment.

He, at first, listed the degrees of negligence of all persons associated in any way in the entire flight(s) and then decided there had been two flights—one from Salt Lake City to Coronado, where the plane landed successfully, and one from Coronado to the crash site—and all negligence prior to that landing was not a proximate cause of the later crash. He decided all causation for the crash was directly related to negligence of the various parties involved with that second flight from Coronado to the crash site. He concluded that the crash occurred directly as a result of Alan Charlesworth attempting to make a 180° turn to reverse course, while flying low over deep snow on the ground; he had become disoriented by whiteout conditions, lost altitude while making the turn, and crashed. He found that Charlesworth would not have been reversing course there but for the causative negligence of FAA employees' failure to adequately provide him with proper information during his approach to Roswell. He determined that the actions of the pilot were 60% causative of the crash, while 40% were due to the FAA employees. Financial experts, through their written testimony, testified as to the financial losses of the heirs of the deceased men. The judge ruled 40% of these be paid to the heirs. These amounted to the following:

Webb heirs: $310,608;
Riding heirs: $405,603; and
Charlesworth heirs: $828,996.

The crash would never have occurred had adequate communications taken place at several stages of the flight. This case illustrates the importance of correct information being provided between the pilot and the briefers and controllers. Charlesworth failed to ask the right questions, the briefers gave him misinformation, and the controllers failed to understand the meaning of the reported weather.

# CHAPTER NINE

# NEW ENGLAND STORM OF JULY 10, 1989

*Mohawk and Black Rock*

## THE EVENT

Severe weather began in upper New York State on the early morning of July 10, 1989. It swept southeast at over 50 mph during the day, severely impacting Connecticut and Massachusetts before reaching eastern Long Island in the late evening. A squall line developed ahead of an eastward-moving cold front near Albany, New York, at about 5:00 a.m. that morning. Several thunderstorms along the squall line grew rapidly in intensity in the early morning hours as they moved southeastward across western New England. A massive, rotating supercell over 11 miles tall became the dominant feature of the storm system and produced a sequence of severe thunderstorms and tornadoes throughout the day and into the evening. The track of high winds and tornadoes across Connecticut during the day is shown in Figure 9.1.

Along the path of destruction, damages exceeded $130 million, 150 people were injured, 1 woman was killed by lightning, and a 13-year-old girl was crushed to death by a falling tree in Black Rock State Park. A total of 17 tornadoes occurred during the day, two of which were of strength F4 on the Fujita tornado scale (winds of 207–260 mph). All others were of strength of F2 (113–157 mph) or weaker and most were short-lived. Torrential rains with as much as 2 inches in 30 minutes fell from thunderstorms. There were hailstorms with hail up to 2.5 inches in diameter. Thunderstorm winds in excess of 90 mph were measured. In the early evening, by far the most destructive tornado struck Hamden, Connecticut. On the ground for only seven minutes, it devastated 30 blocks of Hamden, destroying or severely damaging 400 structures and leveling large industrial areas. Over 90,000 people were left without power—some for over a week. The Mohawk Mountain Ski Area in Cornwall, Connecticut, was virtually destroyed by one of the F4 tornadoes, suffering $1.5 million in damages. This was one of the greatest, most devastating and longest-lasting weather outbreaks of severe windstorms and tornadoes in the recorded history of New England.

Thunderstorms range from isolated, small, local, relatively benign "air mass thunderstorms" to massive, potentially destructive squall lines, containing multiple thunderstorms, that can produce a variety of severe weather events including flash floods (number one killer), lightning (number two killer), hail, downbursts, tornadoes, and derechos (which are straight-line damaging wind events). In its early developmental stages, a single cumulonimbus cloud contains a strong convective updraft of relatively warm and moist air. Electrical charges separate with positive charges in the cloud and negative charges in the ground beneath. When these charges become sufficient for an electrical arc (lightning) to leap between them, a thunderstorm exists. The lightning discharge heats the immediately adjacent air to 10,000°C, causing rapid expansion, which is followed by rapid compression causing the boom known as thunder. Once precipitation (created by the condensation of water vapor in the rising air of the updraft in the cloud) begins, evaporational cooling causes cold air to descend in a downdraft accompanying the falling precipitation. This cold air on reaching the ground spreads out in a cool "gust front" ahead of the storm. The greater the rise of warm air and descent of cool air in the thunderstorm, the greater the likelihood of producing severe weather in the form of damaging winds. These damaging winds take many forms, such as the following:

- a downburst, "an area of strong, often damaging winds produced by a convective downdraft over an area from less than 1 to 10 km [0.6 to 6 miles]" (Glickman 2000, p. 233);
- a microburst, a "downburst that covers an area less than 4 km along a side with peak winds that last 2–5 minutes" (Glickman 2000, p. 491);
- a tornado, "a violently rotating column of air, in contact with the ground, either pendant from a cumuliform cloud or underneath a cumuliform cloud, and often (but not always) visible as a funnel cloud" (Glickman 2000, p. 781);

- a derecho, "a widespread convectively induced straight-line wind-storm"; and
- thunderstorms, which are "a local storm, invariably produced by a cumulonimbus cloud and always accompanied by lightning and thunder, usually with strong gusts of wind, heavy rain, and sometimes with hail" (Glickman 2000, p. 776).

Tornadoes are violent, rotary windstorms associated with thunderstorms. A spiraling column of rapidly rotating air, often appearing as a narrow funnel extending below the thunderstorm, is known as a funnel cloud if it remains aloft and as a tornado if it touches the ground. They are small in size compared with the thunderstorm cloud but can contain such violent winds that they cause near-total destruction of the area they strike.

## LITIGATION

Climatological Consulting Corporation became involved in two lawsuits relating to the impacts of the storms of July 10, 1989. In the case of *Garrell/ Bike v. State of Connecticut*, two sisters, Jennifer and Melanie Bike, participating in a church camp outing were caught in the storm. Guided by employees of a state park, they sought refuge with others in a nylon tent near the park entrance. Storm winds of a microburst blew a large oak onto the tent, killing Jennifer and seriously injuring Melanie. The family sued the state alleging that there was a U.S. government severe thunderstorm warning in effect at the time and the state had failed in its duty to warn the campers of the impending storm. The plaintiff attorneys engaged the services of a highly qualified forensic meteorologist who produced copies of the warnings issued by the National Weather Service, which did include one for that area in effect at the time. The state's defense attorney, Charles Benson, sought the services of Climatological Consulting Corporation.

The staff at the state park stated they had never received the severe thunderstorm warning. A search of their files and those of several other Connecticut state offices failed to locate copies of the thunderstorm warning. Further detailed research indicated that either an operator's error or a computer glitch in the National Weather Service's automated field operations system had inadvertently deleted the necessary address symbols from the transmitted message and several Connecticut addresses, including the state park, failed to receive the warning message. At first, this appeared to absolve the park officials of negligence, but it was soon discovered that park officials had other serious problems. The plaintiffs were able to demonstrate that there were no trained park rangers on duty at the time, that the campers were being supervised by student employees unfamiliar with their responsibility to keep visitors out of harm's way, and that there were nearby concrete bath houses that would have provided safe shelter from an approaching storm whose ominous clouds were clearly visible at the time. The parties agreed to a $4.5 million, out-of-court settlement.

In the case of *Mohawk Mountain Ski Area v. American Home Assurance Company*, the Mohawk Mountain Ski Resort in Litchfield County, Connecticut, brought suit against American Home Assurance Company over the $1.5 million of damages caused by a tornado at the resort. A tornado destroyed the lodge and severely damaged the ski lifts. These were insured by the American Home Assurance Company with a $650,000 policy per event. The damages exceeded $1.5 million. Armed by correspondence from a government meteorologist indicating there might have been two tornadoes, the owner filed two claims for two events. Believing the damages were due to a single tornado, the insurers paid $650,000 for a single event. The owners sued for payment of the claimed second event and reimbursement for their total loss.

The attorney for the insurance company engaged Climatological Consulting Corporation to testify regarding the number of tornadoes impacting the ski area on July 10, 1989, and sought the testimony of Dr. Fujita, the acknowledged world authority on tornadoes. Once again, Dr. Fujita agreed to be a non-testifying consultant. Working together, we analyzed in detail all the data from four National Weather Service weather radars, a sequence of weather satellite images including the 11-mile-high cloud tops of the rotating supercell that produced the most violent weather, and a sequence of low-level aerial photographs of the fallen trees that were taken soon after the tornado impacted the ski slope and the lodge. Such photos provided essential evidence of the treefall pattern, showing the time sequence of the winds that felled the trees. The case was tried in the Superior Court of Litchfield, Connecticut.

Experts engaged by the lodge owners offered their opinion that ground evidence led them to the opinion there had been two tornadoes affecting the lodge, each of which caused half of the damage. Employed by the attorney for the insurance company, assisted by consulting with Dr. Fujita, and examining the wind damage in great detail, I testified there was only one tornado inflicting damage on the ski area. The court ruled in favor of the lodge owners. The insurers considered an appeal but decided the adverse publicity relating to that action would be damaging to their future business and paid the coverage for the alleged, though nonexistent, second event.

This complex storm system likely generated a number of litigation cases. The two I was involved in were quite dissimilar and were interpreted differently by multiple experts.

# CHAPTER TEN

# DOWNING V. BOWATER

*Like Driving into a Marshmallow, Interstate 75, 1990*

## THE EVENT

At 9:10 a.m. on December 11, 1990, a 99 vehicle pileup on Interstate 75 at mile marker 36, near Calhoun, Tennessee, resulted in the deaths of 12 people and injuries to 50 others. The chain of accidents occurred in the dense fog unexpectedly encountered by drivers who had been driving at or above the posted 65 mph speed limit in bright sunshine before encountering the fog, which one described as driving into a huge marshmallow. Another driver said it was as though someone threw a large white towel onto the windshield. A third said it descended all at once and you could not see anything.

A southbound tractor semitrailer slowed upon entering the fog. The similar rig behind it failed to slow and struck the rear of the first one. The uninjured drivers exited their vehicles to check for damage. Then an automobile struck the rear of the second truck. Fire erupted and consumed the three vehicles. Meanwhile, a car in the northbound lane entered the fog and slowed. It was immediately struck by a following car. A pickup truck and two automobiles then crashed into the chain reaction pileup. While there were no fatalities or injuries resulting from the first eight collisions, that changed rapidly in the next few moments as 91 more vehicles smashed into those first eight and the lines of smashed vehicles extended out of the dense fog bank in both the northbound and southbound lanes far enough to be seen from approaching vehicles in time for the drivers to stop, ending the carnage within the fog.

## FOG

Fog is composed of water droplets suspended in the air in close proximity to Earth's surface and, by international definition, reduces visibility below one kilometer (0.62 of a mile). The composition of clouds and fog are similar. They differ only in that clouds are above Earth's surface, while fog occurs at the surface. Fogs occur when the temperature and the dewpoint of the air are equal—or nearly so. The dewpoint is the temperature at which an air parcel becomes saturated when cooled at a constant pressure and water vapor content. Further cooling causes water vapor to condense into suspended water droplets.

All fogs (as well as clouds) require the presence of condensation nuclei in the air. Water vapor condenses on these aerosol particles, which may be dust, salt crystals, or combustion products. Such particles are quite abundant in the atmosphere, though their concentrations vary depending on the distance from their sources. They are most abundant near industrial sites, fires, and heavy traffic area (from vehicle exhaust). The most common types of fog are as follows:

- *Radiation fog* occurs when air is cooled by Earth's surface as it cools at night through nocturnal radiation. When the air temperature falls to the dewpoint and the air becomes saturated, further cooling will result in condensation and suspended droplets are created.
- *Advection fog* occurs when moist air is advected (transported) over a cold surface and the air is cooled to or below the dewpoint, resulting in condensation of water vapor into suspended water droplets.
- *Upslope fog* is when air moves upslope and is cooled by expansion in the ensuing lower pressure. When the air cools to or below the dewpoint, water vapor condenses into suspended water droplets.
- *Frontal fog* is when rain falls (either ahead of a warm front or behind a cold front), adding moisture to the air, which raises the dewpoint. When the dewpoint rises to the temperature, the condensation of water vapor into suspended water droplets creates fog.
- *Ground fog* is usually a shallow radiation fog that does not hide more than six-tenths of the sky and does not extend to the base of any clouds above it.

- *Steam fog* (sometimes called "water smoke") is a fog formed when water vapor from a warm-water surface is added to air above the warm water, thereby raising the dewpoint of the air to the air's temperatures resulting in condensation of some water vapor into suspended water droplets. Steam fog is commonly observed over lakes and streams on cold mornings and, similarly to industrial fog, appears to emanate from industrial smokestacks whenever the industrial exhaust is warm and has a high moisture content.
- *Industrial fog* is fog in the immediate vicinity of industrial plants that create and emit large quantities of water vapor into the surrounding air either in their exhaust stacks or by evaporation from warm-water cooling ponds. The emitted water vapor raises the dewpoint of the air into which it flows. When that dewpoint rises to the ambient air temperature, saturation occurs. Any further water vapor results in condensation with the suspended water droplets creating fog. Industrial fogs occur downwind of the moisture-generating sites. They are "wet fogs" composed of large water droplets that fall slowly toward the ground.
- *Mixing fog* is a less well-known type of fog in which two saturated or nearly saturated air masses with differing temperatures are mixed, resulting in an air mass with an intermediate temperature that is "super saturated," leading to the rapid condensation of very large numbers of very small water droplets. This occurs because the quantity of invisible water vapor contained in the warmer air mass is far greater than that in the cooler air. When they are mixed to a new intermediate temperature, the mixed air is unable to hold the surplus water vapor. This is the fog type that occurred on I-75 near the Hiwassee River on December 11, 1990.

## NTSB REPORT

Nearly two years later, on September 28, 1992, the NTSB adopted Highway Accident Report HAR-92/02, which contained their determination that "the probable cause of the multiple-vehicle collisions on I-75 near Calhoun, Tennessee, was drivers responding to the sudden loss of visibility by operating their vehicles at significantly varying speeds" (NTSB 1992, p. v). The NTSB did not address the question of the cause of the abrupt onset of the extremely dense fog 90 minutes after sunrise on a brilliantly clear and cold morning at the low point of the interstate highway, just north of the Hiwassee River. It did, however, in 7 of its 11 recommendations include the importance of early detection of fog and other visibility-reducing conditions, controlling traffic speed, assuring uniform driver response through educational and training programs, and creation of adequate signage to assure reduced and uniform vehicle speeds in fog conditions. The other recommendations dealt with driver education, transporting hazardous materials, and firefighting.

## PRIOR ACCIDENTS

It was not the first dense fog-related chain reaction series of collisions at that location. Since the interstate opened in December 1973 in the Calhoun area, there had been six multivehicle accidents and 300 other accidents occurring along the short, two-mile segment of the highway just north of the bridge crossing the Hiwassee River prior to December 11, 1990. The prior multivehicle accidents are shown in Table 10.1.

**TABLE 10.1.** Multivehicle accidents in the Calhoun, TN, area prior to December 11, 1990.

| Date | Vehicles | Deaths | Injuries |
|---|---|---|---|
| March 9, 1974 | 18 | 3 | 10 |
| April 19, 1974 | 9 | 0 | 8 |
| June 12, 1976 | 4 | 0 | 0 |
| December 14, 1977 | 14 | 0 | 7 |
| November 5, 1978 | 65 | 0 | 46 |
| April 15, 1979 | 18 | 3 | 14 |
| **Total** | **128** | **6** | **85** |

After these accidents, the Tennessee Highway Department erected fog warning signs with flashers: one for the southbound lanes slightly north of the Exit 36 interchange and one for the northbound lanes shortly south of the Hiwassee River bridge. Both were equipped with flashers (flashing lights) activated by manually operated electrical switches. The Tennessee Highway Patrol was to survey the three-mile stretch of the interstate daily in the early morning and activate the flashers if they observed fog or it appeared likely. On December 11, 1990, a patrolman drove through the fog prone area about 8:00 a.m. He did not see any fog, did not activate the flashers, and drove on to Chattanooga. The fog formed about an hour later and caused the worst of these fog-related accidents (see Table 10.2).

**TABLE 10.2.** The December 11, 1990 accident as compared to the total of all accidents at the location.

| Date | Vehicles | Deaths | Injuries |
|------|----------|--------|----------|
| December 11, 1990 | 99 | 12 | 50 |
| **Total** | **227** | **18** | **135** |

All these accidents occurred between November and June on cold, bright early mornings. All involved the abrupt onset of extremely dense fog about 90 minutes after sunrise.

## LITIGATION

About two weeks after the December 11, 1990, accidents, Mrs. Evelyn Piper Downing, whose son Craig Piper had been killed in the pileup, called Attorney Doug J. Fees of the Huntsville, Alabama, law firm of Morris, Smith, Siniart, Cloud, and Fees (representing plaintiffs) to see if he could represent her in a lawsuit against Bowater Southern Corporation, which operated a huge paper mill at Calhoun, Tennessee, about two miles east of the interchange on I-75 where the accidents occurred. Doug Fees and a law partner, Joe Cloud, drove to Columbus, Georgia, to interview Mrs. Downing. Soon after that they investigated the site and noticed the 300 acres of four wastewater treatment ponds west of the Bowater paper mill, one of which straddled the highway near where the accidents took place. They became aware that three days after the crashes, a Knoxville, Tennessee, newspaper mentioned the similarity of the events to those of the April 1979 crash at the same location. The article discussed a resultant lawsuit brought by the survivors of a woman and her son who were killed in the 1979 accidents. The lawsuit was against Bowater for allegedly releasing millions of gallons of water vapor daily and charged that under the right weather conditions that water vapor caused dangerous fog to form over the interstate. The article mentioned that Bowater's lawyers had settled out of court for undisclosed millions of dollars three days before the case was scheduled to go to trial. Attorney Fees and his firm felt that this was the route to pursue, accepted Mrs. Downing as a client, and went to work both developing a case against Bowater and other potential defendants and began seeking additional clients among the survivors of others killed or injured in the crashes. The case, known as *Evelyn Piper Downing v. Bowater Incorporated* was assigned to Circuit Court Judge Samuel H. Payne in the Eleventh Judicial Circuit of Tennessee in Chattanooga, Tennessee, a short distance southwest of the accident site in Calhoun, Tennessee.

An early legal maneuver by Attorney Fees and his partners was to attempt to have the case become a "class action suit," which is a suit brought by one or more members of a large group on behalf of all members of the group. This would assure every driver involved in the multiple accidents (or their survivors) would be a member of the class, that the one year statute of limitation would be extended, and that all members of the group would receive compensation if there were to be damage awards, rather than only those who filed individually. Certification of the case as a class action suit would provide Attorney Fees with added time to prepare a case, assure him of having all potential plaintiffs as his clients, and provide an opportunity to all those involved in the accident to be compensated if the Bowater mill was found responsible for creating the fog and if damages were awarded to the plaintiffs. Judge Payne denied the motion for a class action trial but permitted individual suits to be filed by any plaintiff attorney within the one-year statute of limitations to be consolidated. The cases would be separate, but all hearings, pretrial proceedings, depositions, and the trial would be handled collectively and simultaneously.

## THE DEFENSE WEATHER TEAM

Bowater, as the defendant, would have to develop a team of consultants, experts, and potential testifying witnesses to defend the firm against all allegations of wrongdoing or neglect. Since weather was evidently a major factor in the development of the fog, they needed a meteorologist—or a team of weather experts—to determine and testify as to the cause of the fog. Harvey Templeton was the in-house attorney at the Bowater mill in Calhoun and was backed up by attorneys at Bowater headquarters in Darien, Connecticut, and legal firms in Chattanooga and Knoxville, and Philadelphia, Pennsylvania. Attorney Templeton regularly sought legal assistance from the firm of Fleissner, Cooper, and Marcus of Chattanooga and the firm of Hodges, Doughty, and Carson in Knoxville, who were further backed up by Steve Phillips of Pepper, Hamilton, and Sheets of Philadelphia.

In their search for a testifying weather expert, Harvey Templeton and Phil Fleissner turned to the Technical Advisory Services for Attorneys (TASA), a firm that maintains a list of qualified experts in many disciplines and arranges, for a percentage of their time and expense charge, to provide their services to attorneys. TASA had previously verified the meteorological expertise available from the staff and associates of Climatological Consulting Corporation and entered into a contract authorizing them to recommend our services when appropriate. Soon after the suit was filed in Judge Payne's Court in Chattanooga, I received a call from TASA advising that Phil Fleissner wished to engage my firm. I soon received a call directly from Phil Fleissner requesting that I meet with him and Harvey Templeton at the Bowater paper mill in Calhoun. I agreed.

At Harvey Templeton's office in the paper mill, a little over two miles southeast of the interchange where the accidents all occurred, he and Phil Fleissner briefed me on the events of December 11, 1990. He told me of the allegations of Bowater's responsibility for creating the fog (or contribution to its density), described the history of the six prior accidents and the allegations of Bowater's possible causation or intensification of those fogs, provided a tour of the plant and the four cooling ponds, showed me the locations of the Olin Corporation chemical plant (which provided chlorine to Bowater) and AKZO Salt Co. industrial installations across the river, and described their relation to the paper mill's processing from timber input to paper output. At the end of a very long day, I left my car at the mill and rode with Phil Fleissner to Chattanooga where I spent two days meeting with his staff and partners and was briefed on their views of the pending litigation and queried about "industrial fogs." They asked if my staff and associates could undertake a comprehensive study of fog occurrences and causation in the Hiwassee River valley with a focus on the I-75 Highway corridor north of the river. I agreed we could, advised that it would be objective and impartial, and that I would likely seek the collaboration of other consulting meteorologists who had special expertise in fog studies and involve CCC associates with specialties such as satellite meteorology. Initially, we would be engaged solely as consultants. If our work appeared potentially beneficial to Bowater's defense, we might later be identified as potential testifying expert witnesses. That would make our work subject to "discovery" by the plaintiffs and us likely to be deposed. Additionally, on the drive back to Asheville, I received a cell phone call from another plaintiff attorney who had a client seeking to sue Bowater in the potential consolidated cases. My declining his request alerted him and the growing group of plaintiff attorneys of the likelihood that my firm had been retained by Bowater as a consultant or potential source of defense witnesses. This gave the group of plaintiff attorneys an early opportunity to investigate our qualifications, cases in which we had been involved, and prior courtroom testimony long before we were identified as defense witnesses.

It became clearly evident that collaboration with a meteorologist highly specialized in abrupt, early morning fog formation in low-ridge and valley topography on clear, cold, winter-season, early mornings would be essential. A search for such colleagues among the members of professional societies such as the American Meteorological Society and the National Council of Industrial Meteorologists led to the awareness that Dr. George McVehil of McVehil–Monnett Associates in Colorado was a leading expert in such occurrences. Dr. McVehil had performed stellar research on fogs in similar terrains in central New York State and had published authoritative peer-reviewed articles about such fogs in the scientific literature. At a second meeting in Chattanooga, I briefed Phil Fleissner and Harvey Templeton on the results of a preliminary review of the archived weather data pertinent to all the I-75 fog-related accidents in southeastern Tennessee, introduced Dr. McVehil, and outlined a proposed work program to determine the cause of the December 11 fog on I-75 near Calhoun. The attorneys were impressed with Dr. McVehil and agreed to engage him directly, with the understanding he would collaborate fully with the team assembled by Climatological Consulting Corporation. The CCC team would consist of staff and associates of the company, including Walter Bohan in Chicago who would utilize the satellite weather data archives at the University of Wisconsin–Madison to analyze the locations, extent, and timing of fog over a multistate area on December 11, 1990, in relation to topography.

## PLAINTIFF ALLEGATIONS

There is widespread general awareness that paper mills (and many other industrial activities) generate large volumes of water vapor that contribute to fog formation in the vicinity of the plants. Doug Fees shared this awareness and researched references mentioning fogs in the vicinity of paper mills, especially those impacting highway travel. These included the following:

- Champion Paper Mill in Canton, North Carolina, immediately adjacent to Interstate 40 where several fog-related accidents led to litigation;
- Three industrial plants in Green Bay, Wisconsin, adjacent to Tower Drive Bridge, where several highway accident deaths occurred;
- Westvaco Paper Mill in North Charleston, South Carolina, where a planned, downwind interstate highway was rerouted to avoid frequent industrial fog; and
- Weyerhaeuser's huge mill in Kamloops, Oregon, where (during a plant expansion) the company installed a $7 million steam pipeline several

hundred feet up a hillside to emit their water vapor well above the frequent, near-surface, and 500–1,000-foot levels of atmosphere temperature inversions.*

Attorney Fees believed the multiple industries, Bowater, AKZO, and Olin, on the banks of the Hiwassee River east of I-75 all caused or contributed to the fog on the interstate, and they became the targets of the litigation. He alleged that these industries had known for 20 or 30 years that they should have implemented control technologies to minimize fog. The industry officials stated they did not believe their industrial activities either caused or contributed to the fogs that they believed were "natural" and had been prevalent long before the industries were located on the banks of the Hiwassee River.

## EARLY STUDIES

After the November 5, 1978, accident (fifth set of accidents), with 63 vehicles and 46 injuries, but no deaths, the Highway Safety Planning Division of the Tennessee Department of Transportation became concerned that the several fog-related accidents in the same location might have been either caused or aggravated by the Bowater paper mill and/or the several industries associated with the mill. They contacted the University of Tennessee Transportation Center to propose a research study to determine the cause of the repeated, early morning, extremely dense fogs at the same portion of I-75. The study was assigned to Dr. Wayne Davis, an assistant professor in the department of civil engineering. Dr. Davis obtained the detailed weather data relevant to the five accidents (1974–78) from the National Climatic Data Center in Asheville, North Carolina. He soon discovered remarkable similarities in the weather related to each of the accident events. All occurred shortly after sunrise on clear, cold, calm mornings after an especially warm day. He

---

* "Temperature inversion: A layer in which temperature increases with altitude. The principal characteristic of an inversion layer is its marked static stability, so that very little turbulent exchange can occur within it. Strong wind shears often occur across inversion layers, and abrupt changes in concentrations of atmospheric particulates and atmospheric water vapor may be encountered on ascending through the inversion. When an inversion is mentioned in meteorological literature and discussion, a temperature inversion is usually meant. See frontal inversion, subsidence inversion, tradewind inversion" (AMS 2012c).

believed this indicated a high likelihood that a strong temperature inversion was present at a few hundred feet above ground level. Such an inversion, with very light winds in the air below it, would "trap" cold air, pollutants, and water vapor beneath it, providing the essential ingredients for fog. He decided to take aerial photographs of the area whenever he foresaw similar weather conditions to be likely. Three early morning flights in December of 1978 showed extensive fogs covering the area of the I-75 exchange at Exit 36 and the industrial plants including Bowater. He felt the Bowater mill might have caused the fogs. His report stated so.

The officials at the Tennessee Department of Transportation sent a copy of Dr. Davis' report to the president of Bowater and requested a meeting at which Dr. Davis could present the results. The meeting was held in January 1979. Bowater officials believed Dr. Davis' conclusions were "unfounded." Then the April 15, 1979, accidents involving 18 vehicles, 3 deaths, and 14 injuries occurred. Dr. Davis proposed a more comprehensive study, which was not approved. In January of 1980, the Tennessee Department of Transportation Office of Research contacted Dr. Steven Hanna at the Atmospheric Turbulence and Diffusion Laboratory of the National Oceanic and Atmospheric Administration in Oak Ridge, Tennessee, and invited him to make a proposal for a comprehensive study of fogs near Exit 36 of I-75 in Calhoun. He proposed a major study. The proposal was scaled down; then Dr. Hanna worked on the smaller study between August 1980 and May 1981, which concluded the following:

- "heavy fog . . . is three times more likely at the I-75 location in Calhoun than at Knoxville, Chattanooga, or Oak Ridge," and
- "both the river and the mill were sources of large amounts of water vapor, which is critical to the development of fog," but
- "the contribution of the mill to the fog was difficult to determine," and
- "a more comprehensive study would be required to resolve the uncertainties" (Weiss 2004, p. 37).

No further study was made until after the December 11, 1990, accidents.

Apparently concerned that the Davis and Hanna studies might somehow implicate Bowater's cooling ponds or stack effluents as potential sources of water vapor for the highway fogs, Bowater engaged Environmental Research and Technology, Inc. (ERT), in 1979 to conduct a study of the possible causes of the accident-related fogs, referred the Davis study to ERT, and sought recommendations on how its conclusions might be scientifically refuted.

That study, "A Preliminary Review of Fog Occurrences in the Vicinity of Calhoun, Tennessee" (Weiss 2004, p. 29), indicated that perhaps as much as 40% of the water vapor related to such fogs might have come from Bowater's plant. Since it was privately funded by Bowater, focused primarily on water vapor potential sources, did not relate them to specific events, and suggested Bowater created some of the water vapor, it was retained in Bowater's files and not made public for nearly 14 years. After the 1990 accidents and the onset of the resultant litigation, it became known as the "secret study" and remained so until 1992 when the Tennessee Supreme Court ruled it could be made available to the plaintiff's attorneys in the *Downing v. Bowater* litigation over the December 11, 1990, accidents.

## THE ANALYSIS

Once engaged by Phil Fleissner's law firm on behalf of Bowater in early 1991, Dr. McVehil and I—with the help of our staff and associates—began a thorough study of the physiography and meteorology relevant to the abrupt onset of extremely dense early morning fog. Just as in the case of a pathologist performing an autopsy, it is essential that a forensic meteorologist examine every aspect of the weather relating to the event, its prior occurrences, the research and findings of all who had examined prior events, as well as every bit of information available before venturing any opinions on causation. The coincidence of complex topography, the placement of the interstate, time of day, and the sequence of several meteorological factors appeared to be intimately involved in the abrupt formation of an extremely dense mixing fog shortly after sunrise near the interstate of I-75 and Lamontville Road on December 11, 1990.

## TOPOGRAPHY

The complex topography and the relationship of rivers, lakes, ridges, valleys, highways, industrial sites, three local weather stations, and the multiple accident sites are illustrated in Figure 10.1. The interstate runs generally northeasterly–southwesterly along the bottoms of a series of similarly oriented, shallow valleys, separated by generally 300-foot-high ridges. These are characteristic of the southern Appalachian Mountains throughout eastern Tennessee. The lower reaches of the Hiwassee River cut across the ridges and valleys as the river flows west-northwestward into the eastern extremity of Lake Chickamauga, just west of the bridge where Interstate 75 crosses the river. The Hiwassee River flows rapidly westward from the mountains of western North Carolina, but it slows dramatically before reaching Calhoun. The slope of the river is seven-tenths of one foot per mile upstream of the Hiwassee River bridge and flat west of the bridge as the river joins Lake Chickamauga. The 35,400-acre Lake Chickamauga was created by the building of the Tennessee Valley Authority's (TVA) fourth dam in 1940. The 120-foot high dam, 36 miles northeast of Chattanooga, backed up the waters of the Tennessee and Hiwassee Rivers to create the lake terminating just west of the I-75 bridge across the Hiwassee River.

The Bowater paper mill is located on the north bank of the Hiwassee River, two and a half miles upriver [east-southeast (ESE)] of the bridge. The interstate—when traveling southbound—goes from the center of the moderately forested and narrow North Mouse Creek valley (upper right of Figure 10.1), which for the two miles up valley from the accidents site slopes downward toward the south-southwest at only eight feet per mile. Interstate 75 then curves briefly toward the southwest, past the open mouth of the southern end of the broader and more open agricultural Spring Creek valley (upper, more central valley in Figure 10.1), which has a slope downward toward the interstate of 27 feet per mile. The two valleys are separated by a 300-foot ridge. The accidents occurred where I-75 reaches its lowest point, curves again to the south-southwest, and climbs slowly uphill onto the Hiwassee River bridge and the higher ground in South Mouse Creek valley south of the Hiwassee River bridge.


## THE WEATHER

The weather on December 10 and 11, 1990, bore remarkable similarity to that of the six prior events as well as that on the dates of the two (of three) dates investigated by Dr. Wayne Davis for the Transportation Center of the University of Tennessee. These eight prior fogs all occurred shortly after sunrise on clear (or nearly clear), initially calm, cool (or cold) mornings following very warm days. Near-surface temperature inversions were present in all cases. Upper winds (in the first 1,000 feet above the land surface) came from directions between southwest and northwest, with several slowly shifting from coming from the southwest to coming from the northwest (i.e., blowing from the accident area southeast toward the paper mill) at the time of the usually abrupt onset of the fog. On December 10–11, 1990, the weather was dominated by a very slowly eastward-moving fair weather high pressure area both at the surface and in the lower several thousand feet of the atmosphere. The "ridge line" (region of highest pressure) of this meteorological "fair weather high" moved slowly eastward over Calhoun at about 7:00 a.m. on December 11 (see Figure 10.2). Regionally, the very light, upper-air (500–1,000 feet above ground) winds shifted, originally coming from the northwest and then gradually changing to a flow from the southwest.

There are three weather stations on the Bowater property. One (called Bonorth) is half of a mile north of the mill; a second (Bosouth) is three-quarters of a mile south-southwest of the mill; and the third (Bomet) is one and three-quarters of a mile northwest of the mill and half a mile south-southeast of the site of the accidents. During the early morning hours (prior to and at the accident's time), the winds recorded at the three surface weather observing stations were as follows:

- at Bonorth, coming from the northeast at 1.8 mph;
- at Bosouth, coming from the south-southeast at 1.3 mph; and
- at Bomet, coming from the northwest at 1.2 mph.

They indicated a very slow inflow of surface air toward the mill to replace the air slowly rising above the mill because of the heat of the paper-making process. Otherwise, the winds were calm except for very slow "down valley" surface cold air drainage winds in Spring Creek and North and South Mouse Creek valleys. Air draining out of North Mouse Creek valley flowed across the interstate and past Bomet toward the mill. There were no winds, however light, at either the surface or aloft flowing from the mill toward the interstate.

On December 10, the daytime temperatures at the surface rose to a daily maximum at 4:00 p.m. of 68°F in Chattanooga, 62°F in Knoxville, and 62.4°F at the Bomet station between Exit 36 and the paper mill. At the time of the 4:00 p.m. maximum temperatures, the dewpoint temperatures were 31°F in Chattanooga, 32°F in Knoxville, and 32.4°F at the Bomet station. Figure 10.3 shows the hour by hour temperature in Chattanooga, Knoxville, and at Bomet from noon on December 10 to noon on December 11. A horizontal line on the graph represents 32°F, the freezing temperature. The temperature at these three locations fell from the December 10 afternoon mid-60s highs to early morning lows on the morning of December 11 of 31°F at Chattanooga, 33°F at Knoxville, and 29.8°F at Bomet between 7:00 and 8:00 a.m. They then rose rapidly during the hour after sunrise. The more conservative dewpoint temperatures were far more constant until water vapor was removed from the air by the deposition of dew and frost on the land surface during the night, causing the dewpoints to lower as moisture was removed from the air.

Light frost occurred at Chattanooga from 5:00 to 9:00 a.m., while temperatures and dewpoints were below freezing. Some dew occurred at Knoxville between 9:00 p.m. on the 10th and 9:00 a.m. on the 11th, while the dewpoint was slightly depressed by the condensation of water vapor in the air to create dew at the ground. Some light, early morning radiation fog existed briefly in low-lying areas near the airport at about 7:00 a.m. When the sun rose at 7:39 a.m. on December 11 in Calhoun, Tennessee, the valley bottoms and the ground at the Lamontville Road interchange sparkled with the ice crystals of what spectators described as a very heavy frost. At the ridge tops, there was a heavy dew.

The cold air at the bottom of North Mouse Creek was only 100–200 feet deep. At the tops of the 300-foot ridges separating the valleys, the air temperature was above 50°F. At 8:00 a.m. on December 11, a farmer, attempting to start a dew-soaked tractor on the ridge between Spring Creek and North Mouse Creek valleys, noted the temperature to be 54°F. The warming of Earth's surface during the next 40 minutes caused the temperature to rise 1.5°F and the dewpoint to rise 2.5°F at Bomet, as the heavy frost was rapidly converted to water vapor in the air immediately above the ground. This process increased more rapidly between 9:00 and 10:00 a.m. as melting frost and evaporating dew fed huge quantities of water vapor into the air, making (and keeping) it saturated.

Shortly after 9:00 a.m., the heating of the sun on the surface of the land created a gentle rising of the air immediately above the surface, mixing it with the air of the 300-foot ridge top levels. The valley bottom air (saturated) at 28°F and the ridge top level air (also saturated) at 54°F mixed and became "supersaturated air" at an intermediate temperature. If equal quantities mixed, the temperature of the mixed air would be 41°F. Figure 10.4 illustrates the nonlinear relation of the water vapor capacity of air at differing temperatures. The state of the air can be represented by a point at any temperature and water vapor content.

In the diagram, provided by Dr. McVehil, temperature increases from left to right, while water vapor content increases from bottom to top. The solid curved line represents the saturation of the air in terms of grams of invisible water vapor the air can hold at various temperatures. To the right of that line, the air is unsaturated. To the left of it, condensation into droplets will occur. When saturated, cool air (point B) mixes with saturated, warm air (point A), the resulting temperature will be somewhere along the line A–B, which is to the left of the curved line, indicating the mixed air will be supersaturated and droplets must form. Reference to Figure 10.4 would show that all points on the line of temperature from 54° to 28°F of saturated air are to the left of the curved line of saturation values. The excess water vapor immediately condenses into billions of tiny, suspended water droplets, as there are ample condensation nuclei available from the exhaust gases of vehicles on the interstate. The resulting droplets are very numerous and so very small, they seem to float in the air (with no "fall speed"). This creates an extremely dense "dry fog," decreasing visibility often to ranges of 10 to 30 feet.

## AERIAL VIDEO AND SLIDES

Two flights were made on the morning of December 11, 1990, to document the scene. The first, conducted by Johnny Miles, began at 10:15 a.m. (about one hour after the accident sequence began). Mr. Miles flew in a series of wide circles while videotaping the scene below. The second flight began about 11:00 a.m. It was flown by Mr. Torbitt, employed by Bowater. He took over 100 color slides. We studied all the images from both flights. Mr. Miles' video contained 30 frames per second for over 15 minutes, providing over 30,000 images. The pertinent ones showed that approximately one hour after the accidents:

- the dense black smoke from the burning vehicles rose several hundred feet into the atmosphere and drifted toward the northeast at all levels;
- the northeastern edge of the fog was about one mile northeast of the interchange;
- the steam plumes from the Bowater stacks were east of the fog and both leaned and traveled to the northeast about two miles east of the interstate;
- the fog did not extend above the ridge separating the Spring Creek and North Mouse Creek valleys;
- the fog filled the lower part of North Mouse Creek valley and extended southeastward across the interstate but did not reach the Bowater mill; and
- there were separate areas of fog south of the Hiwassee River and in South Mouse Creek valley.

Figure 10.5 shows a reproduction of frame 29 at 5 minutes and 19 seconds into the aerial video on the morning of December 11, 1990. The aircraft was southwest of the accident, headed southeast. The image shows the smoke from the burning vehicles on the interstate. The steam plumes from the Bowater stacks are in line with the left wing about two and a half miles southeast of the accident. The direction of movement of these plumes is toward the northeast (050 degrees). The smoke from the burning vehicles is also drifting toward the northeast, and the ridge line between the valleys to the northwest extends above the fog, indicating the fog is less than 300 feet in depth.

## REPORT TO DEFENSE COUNSEL

In April 1993, we completed our fieldwork and research and produced "A Report to Counsel on the Causes and Origin of Fog at Exit 36 on Interstate Highway 75 on December 11, 1990" (Haggard and McVehil 1993, 167 pp.). The document was 167 pages in length and contained 6 tables, 85 figures, and 49 technical references. The report

- described the atmospheric elements and processes involved in the formation of fog, the water vapor in the atmosphere, the effects of the terrain–land–water relationships, air stability and temperature, temperature and dewpoint regimes, and condensation nuclei and fog optics as background physical processes;
- contained a discussion of historical weather and fog in east Tennessee;
- described the factors related to industrial sources in fog formation;
- contained a section on satellite meteorology in relation to fog analysis;
- illustrated the climatic history of wind direction and speeds at multiple U.S. locations as compared with those of eastern Tennessee and at the Bowater paper mill in Calhoun, Tennessee;
- reviewed in depth the meteorology of December 9–11, 1990, in relation to regional and local areas;
- discussed an atmospheric moisture budget, including quantitative estimates (under the existing conditions) of natural water areas, industrial ponds, dew and frost deposits, evaporation, and stack emissions; and
- interpreted aerial videos and photos of the fog made during the morning of December 11, 1990, for the interpretation of the location of the fog and the location and direction of the movement of the steam plumes from the Bowater exhaust stacks.

## CONCLUSIONS

Our analysis led to our conclusion that the fog on Interstate 75 at the Lamont-ville Road interchange (Exit 36) at about 9:05–9:30 a.m. on December 11, 1990, was caused by natural conditions and normal atmospheric processes consistent with the laws of physics governing the formation of fog, the history of fog formation in the Bradley and McMinn County area of Tennessee, and the meteorological conditions on December 11, 1990, and that the water vapor from the Bowater facility and/or its wastewater treatment ponds did not make a significant contribution to the fog on Interstate 75 at the Lamontville Road interchange. Dr. Eschenroeder, the plaintiff's weather expert, engaged by Doug Fees, utilized computer modeling to illustrate his concept that water vapor from the waste water ponds and the mill stacks rose into the atmosphere above the surface winds (blowing from the highway toward the plant), moved westward, and descended at the interchange into the cold surface air to create the fog. Contrary to Dr. Eschenroeder's computer analysis and opinion, which were unknown to us at the time, we focused our analysis on the complex topography, the observed weather-related facts, witness statements, aerial photography, and meteorology processes to arrive independently at our contrary opinion.

## DEPOSITIONS

In many cases, when a deposition is taken of a potential testifying expert by the lawyers on either the plaintiff or defense's side of the case, the opposing attorney brings his/her similar expert to assist in preparing deposition questions. Such was not the procedure in *Downing v. Bowater*. Neither Dr. McVehil nor I were invited to Dr. Eschenroeder's deposition. Indeed, we were never privileged to read the transcript of it. He, in turn, was not present when Mr. Fees deposed me.

Shortly after our "Report to Counsel" was completed in April 1993, I was deposed by Attorney Doug Fees, who had presumably read the report. I, accompanied by Defense Attorney Steve Phillips, was deposed in a hotel in downtown Asheville. The room was lined with large photos of smashed vehicles, smoke, fog, and highway carnage, that were frequently pointed out during the questioning. The questions assumed that the water vapor from the fog was largely produced by either the Bowater mill or the four "warm cooling ponds" between the mill and the highway. The deposition took less than a day and appeared to me to be an effort to create my sympathy for the dead and injured more than a probe of the content of our report. We were, however, asked to produce our entire file of data, correspondence, and analyses, together with videos, photographs, eye witness statements, and all our billing documents. All these were copied and provided to the dozen or so attorneys present, together with copies of our report.

## SETTLEMENT

Nearly three years after the collision and after several postponements of the trial date, then set for September 13, 1993, Judge Samuel Payne set the final trial date for January 1994 and advised all parties it was a firm date. By that time, Attorney Fees had negotiated settlements totaling over $2.5 million with AKZO (Salt Co.) and Olin Corporation, with the Tennessee Department of Transportation and the Tennessee Highway Patrol, and with several trucking companies whose trucks had struck and demolished several vehicles in the fog. The sole remaining defendant was Bowater. Doug Fees had not entered into settlement negotiations with them. He was confident that potential jurors would have great sympathy for the families of the dead and injured, would presume Bowater was heavily insured, and would believe with the help of Dr. Eschenroeder's computer modeling and their own perception that paper mills were responsible for creating fog. He felt his 44 clients stood a good chance of generous jury awards. He knew, also, that the adage "a good settlement is better than a bad trial" was true. Bowater's defense team was equally aware of the possibility of a huge jury award based on sympathy. After negotiation, a $7 million settlement to the plaintiffs was agreed on. Bowater closed its cooling ponds between the plant and the interstate.

## HIGHWAY FOG DETECTION AND WARNING

Following the crashes of December 11, 1990 (and in accord with a recommendation of the NTSB), the Tennessee Department of Transportation and the Tennessee Department of Safety designed and implemented a massive, low-visibility warning system. By continually monitoring the system sensor data, the onsite computer detects and predicts conditions conducive to fog formation and detects significant reductions in traffic speed. When established threshold criteria of either are met, the central computer sounds an audible alarm in the Highway Patrol Office. The dispatchers then activate the warning signals and notify troopers. Low visibility caused the highway to be closed twice in the 10 years between 1993 and 2003—once because of fog and once because of smoke from a nearby fire. There has been only one fog-related crash during those years following the implementation of the system.

# CHAPTER ELEVEN

# PALATKA PAPER MILL

*The Other Side of the Coin, Palatka, Florida, 1991*

## THE EVENT

An outbreak of severe weather, including violent thunderstorms, hail, tor-rential rains, and gales, struck Palatka, Florida (see Fig. 11.1), during the morning hours of April 23, 1991. Georgia-Pacific's paper mill was extensively damaged as well as other buildings and power lines in the county.

## THE PAPER MILL

The westward-facing paper mill on the east bank of the wide, northward-flowing St. Johns River was built in 1947. The mill was expanded over the years, and in early April 1991, a contractor was engaged to install some large, new equipment on the second floor in the west side of the mill. The contrac-tor removed a large section of the exterior brick wall on the second floor to provide access to move in the new equipment. To protect the exposed inte-rior space and the massive electrical panels on the east wall of the exposed space, the contractor temporarily erected wooden posts in the opening and mounted heavy-duty plastic sheets to the posts. Between 9:00 and 9:30 a.m. (EDT) on April 23, 1991, thunderstorm winds, rain, and hail tore the plastic from the posts, allowing the rain and hail into the room and against the elec-trical panels on the east wall of the space. The electrical circuits shorted and the mill shut down. The mill managers felt the contractor was responsible for that portion of the damage created by the electrical shorting in the opened room because of inadequate protection from the elements. The contractor argued that the weather was extreme, beyond normal expectations, and that he had provided protection adequate for normally expected weather.

Georgia-Pacific engaged the law firm of David H. Popper & Associates in Orlando, Florida, to investigate the potential liability. The law firm sought to find whether the weather was foreseeable, in which case the contractor might have been liable. Long after the event, they contacted Climatological Consulting Corporation and requested an investigation and report on the weather conditions at the site of the mill and what was foreseeable.

## SITE VISIT

After being contacted by the law firm, arrangements were made for a site visit. I traveled to Palatka and met with the attorneys at the law firm and then went to the mill site where I was shown the functioning of the mill and was introduced to several employees who had been on duty on April 23, 1991, who related firsthand accounts of their experiences on the day of the damage and showed me hundreds of photographs taken on that day. The tour and interviews took two days and provided a wealth of detailed information, including the exposure to west winds flowing over the wide river. I spent a third day visiting the sheriff's office, a Chevrolet dealership, and various locations in the Palatka area that had suffered hurricane-type damage. After the site visit, I proposed a study to the attorney, to which he agreed. To conduct the requested study, my staff associate and I obtained certified copies of the pertinent archived weather data for April 23 from the National Climatic Data Center including hourly and special surface weather, twice-daily upper-air observations, North American surface weather charts, national weather depiction charts, 12-hourly upper-air constant pressure charts, weather satellite photographs in both visible and infrared, radar film images, radar geographic overlays together with hourly radar summary charts and radar observation summaries, and the publication *Storm Data* for April 1991. Lightning data were obtained from the archives of the National Lightning Detection Network.

The surface and upper-air weather charts provided the big picture of the weather patterns over the United States. They included a surface front extending from Brownsville, Texas, to the Atlantic off the Georgia coast. This was a cold front over the western Gulf of Mexico moving slowly south. Over

Florida, south of Pensacola, through Tallahassee, and north of Jacksonville, it was stationary. Early morning temperatures north of the front were in the mid-50s, while those south of it over Florida were 15° to 20°F warmer, being 70°F at Palatka. The upper-air charts showed relatively warm, moist, and somewhat unstable air south of the front with winds from the west increasing with height from 50 mph at 2,000 feet to over 100 mph at 40,000 feet. These winds brought warm, moist, and convectively unstable air from over the Gulf of Mexico, which provided ample moisture to fuel strong convection and thunderstorms over the Florida Peninsula, including the Palatka area. The weather satellite imagery in both visible and infrared (heat) frequencies showed extensive, thunderstorm, high-level (above 60,000 feet) tops with cirrus "blowoffs" moving east at speeds of 40 to 70 mph. The vertical radiosonde soundings of surface to upper-air temperature, humidity, and winds, made over Florida, south of the surface front, showed the conditions were favorable for strong vertical convection, conductive to strong thunderstorm activity, while those north of the front were indicative of cooler and far more stable weather.

The National Weather Service forecast for central and northern Florida on April 23, 1991, included a series of severe thunderstorm warnings between 7:47 and 11:02 a.m. (EDT). They included such statements as the following:

- "This is a dangerous storm. If you are in its path you should prepare for damaging wind in excess of 55 mph . . . large hail and deadly lightning."
- "Reports from Putnam County indicate that marble size hail has fallen in Putnam Hall and Carraway . . ."
- "At 8:55 am EDT storm spotters from Keystone Heights reported a possible tornado touchdown at the airport . . ."
- "At 9:55 am, radar showed a tornado along the Putnam and Flagler County line near Crescent City moving east 40 mph" (NWS 1991).

The U.S. government monthly publication *Storm Data* for April 1991 (NCDC 1991) report on Florida for April 23 included an entry for Putnam County, Putnam Hall, and eastern Palatka stating: "Strong wind gusts blew out store windows, uprooted trees, and toppled power lines and utility sheds. Medium to large hail fell. Most of the damage was at a Chevrolet dealer showroom and an adjacent auto parts store in Palatka. Two people were injured by flying glass at the car dealership. A paper mill east of Palatka suffered damage significant enough to force it to close temporarily" (NCDC 1991, p. 48).

The most valuable tool for tracking the thunderstorm activity over Florida on that day was weather surveillance radar. Thunderstorms are mesoscale meteorological events. Mesometeorology is the study of atmospheric phenomena on a scale smaller than the cyclonic scale (highs, lows, and fronts) but larger than the micrometeorology scale (microbursts and very small scale weather events). Mesometeorology is concerned with the detection and analysis of the state of the atmosphere as it exists between meteorological stations and usually well beyond the range of normal observations from a single point. Radar has proven to be a valuable means of detecting and analyzing these smaller scale but often extremely violent and damaging weather events. Radar and satellites are currently the most widely used tools.

The U.S. National Weather Service operates a network of weather surveillance radar stations capable of detecting and tracking precipitation areas. The network includes weather radars at Apalachicola, Tampa, and Daytona Beach, Florida, and Waycross, Georgia, which are all capable of observing the intensity, vertical extent, and movement of precipitation "echoes" over the northern Florida Peninsula, including the area around Palatka, Florida. Repeater scopes on these radars are routinely filmed, and the film is archived at the National Climatic Data Center in Asheville, North Carolina, together with the observational forms completed hourly (or more frequently) by the observers at the radar stations. The filmed images from these four radars were projected on a geographic overlay for each of the four radars, permitting detailed reconstructions of the geographic placement and analysis of the evaluation of the mescoscale events on April 23, 1991. The Apalachicola weather radar is approximately 175 nautical miles west of Palatka, Florida. The best imagery distance for detailed analysis is usually less than 125 nautical miles. Thus, the Apalachicola weather radar was better suited for analysis of the development, evolution, intensity, and motion of thunderstorms west of Gainesville, rather than over Palatka. The Daytona Beach weather radar is approximately 40 nautical miles southeast of Palatka, an "ideal distance" for analysis. Unfortunately, the radial distance rings electronically portrayed on the repeater scope during filming were difficult to align geographically, making accurate echo reconstruction analysis difficult. The Tampa weather radar is 125 nautical miles south-southwest of Palatka, at the outer limit of optimum range for reconstructive analysis. The Waycross, Georgia, weather radar is 105 miles north-northwest of Palatka, within the optimum range for analysis. Also, the Waycross film was determined to be stable in terms of azimuth and distance calibrations and provided clear images of the thunderstorm activity affecting Putnam County, Florida, and Palatka that morning.

In addition to the filmed images, each of the weather radar observers at these four radar sites recorded "weather radar observations" at specific times, describing in coded detail the significant features on their radars. Times (EDT) of observations especially pertinent to this study are shown in Table 11.1.

**TABLE 11.1.** Weather radar observation times for the northern Florida Peninsula on April 23, 1991.

| Waycross, Georgia | Apalachicola, Florida | Tampa, Florida | Daytona Beach, Florida |
|---|---|---|---|
| 8:00 a.m. | 7:25 a.m. | 7:59 a.m. | 8:08 a.m. |
| 8:26 a.m. | 8:25 a.m. | 8:25 a.m. | 8:28 a.m. |
| 9:00 a.m. | 9:01 a.m. | 8:55 a.m. | 9:05 a.m. |
| 9:35 a.m. | 9:31 a.m. | 9:25 a.m. | 9:31 a.m. |
| 9:59 a.m. | 10:03 a.m. | 9:56 a.m. | 9:51 a.m. |

Each observer noted and tracked extreme thunderstorm activity affecting Alachua and Putnam Counties in Florida before and after 9:00 a.m. (EDT) on April 23, 1991.

The intensity of the radar echoes observed by the strength of the reflected beam from the rainfall in the thunderstorms is coded in six levels of intensity as shown in Table 11.2.

**TABLE 11.2.** Radar Echo Intensity

| Video integrator and processor (VIP) | Intensity symbol | Echo intensity | Rainfall rate (in/hr) |
|---|---|---|---|
| 1 | - | Weak | <0.2 |
| 2 | | Moderate | 0.2–1.1 |
| 3 | + | Strong | 1.1–2.2 |
| 4 | ++ | Very strong | 2.2–4.5 |
| 5 | X | Intense | 4.5–7.1 |
| 6 | XX | Extreme | >7.1 |
| U | U | Unknown | Unknown |

The Daytona Beach and Waycross radar operators coded level six VIP returns crossing Putnam County, Florida, between 8:30 and 9:30 a.m. (EDT) on April 23, 1991. "VIP levels" relate to the intensity of the rainfall reflecting energy back to the radar. A system known as the digital video integrator and processor (DVIP) was developed to meet requirements for the continuous, quantitative data of rainfall intensity (which is related to storm severity). The DVIP automatically produces up to six levels of intensity data, which may be displayed individually or simultaneously on the radar scope. This permits constant monitoring of echo intensities with each rotation of the radar antenna.

Analysis of the encoded data and the film from the four radars indicated a level six (extreme) thunderstorm moved from Putnam Hall to Palatka, Florida, at a speed of movement for the thunderstorm cloud of 36 knots (41 mph), reaching Palatka shortly after 9:00 a.m. (EDT) on April 23, 1991. The severity of a thunderstorm may be indicated by several factors observable on the radar, other than echo VIP intensity. Extremely powerful thunderstorms will build to great heights in the atmosphere. The "radar tops" can be measured by tilting the antenna upward. Thunderstorms whose tops are higher than the tropopause are usually the most severe. The tropopause is the boundary of the atmosphere separating the usually stable stratosphere from the lower troposphere. The troposphere is the lower portion (roughly 10 miles deep) of the atmosphere in which weather characteristically occurs. Ordinary thunderstorms usually exist only in the troposphere. Severe thunderstorms may extend a few thousand feet into the stratosphere. On April 23, 1991, the extreme thunderstorms, which affected Putnam County, Florida, extended over 15,000 feet into the stratosphere, an indicator of the intensity of the updrafts, which "overshot" the 50,000-foot tropopause height by 15,000 feet, extending to 65,000 feet, as measured by the weather surveillance radar at Daytona Beach.

## CONCLUSIONS

In January 1996, we submitted our report. Our 12 opinions were as follows (Haggard and Smutz 1996):

1. On April 23, 1991, Putnam County, Florida, experienced an extremely severe thunderstorm between 9:00 and 9:30 a.m. (EDT).
2. The core of this thunderstorm moved across the county from Putnam Hall to East Palatka and into St. Johns and Flagler Counties.
3. The convection in this thunderstorm was so strong it "overshot" the 50,000-foot tropopause and extended 15,000 feet into the stratosphere.
4. The precipitation radar echo of the storm was VIP level six or "extreme" (the highest reportable value).

5. The thunderstorm was accompanied by hail as it approached Palatka.
6. The storm matured and was "collapsing" on reaching Palatka and East Palatka.
7. Severe downbursts are a frequent accompaniment of such collapsing cells and were present at the time the thunderstorm reached the Georgia-Pacific mill at St. Johns River.
8. Torrential horizontal rain (sheets of windblown rain at a shallow angle) accompanied the outburst winds at the leading (ESE) side of the storm.
9. Atmospheric computations of wind gust at the Georgia-Pacific mill indicated potential speeds in the range of 70–80 mph.
10. Many damage reports were characteristic of wind gusts of 73–112 mph; some damage reports were characteristic of wind gusts of 113–157 mph.
11. Such wind gusts are in the wind speed range of "moderate" and/or "significant" tornadoes and are of hurricane speed.
12. Weather of such severity is an extremely rare event at any specific inland location in northern Florida in April.

The attorneys, after reviewing the report, decided it was unlikely that a court would find the contractor guilty of negligence under the circumstances, and they decided not to sue.

CHAP TE R TWE LVE

# MCNAIR V. USA

*Obstructing Clouds, Gainesville, Florida, 1995*

## THE EVENT

At approximately 10 minutes past midnight on the early morning of June 7, 1995, a six-passenger, single-engine Piper Cherokee aircraft crashed in a swamp approximately 5 miles east-northeast of the Gainesville, Florida, airport, the intended destination. All six persons aboard the aircraft (five passengers and the pilot) were found dead at the scene when the wreckage was located in the sparsely populated area at about 11:30 a.m. that morning.

## LITIGATION

The families of the deceased—after learning the details of the flight, the communications between the pilot and various traffic controllers and Flight Service Station personnel, the recorded weather conditions and those reported to the pilot, and other relevant facts—brought a suit against the U.S. government, claiming the crash was due to negligence by government employees who provided erroneous information to the pilot concerning the weather conditions at the Gainesville, Florida, airport, misleading her to attempt to land under conditions she was neither equipped nor qualified to fly in.

125

# CHRONOLOGY
## June 6, 1995

8:45 p.m. (EDT)
The flight originated at Covington, Georgia, near Atlanta. The pilot, Loveda McNair, planned a visual flight rules flight to Gainesville, Florida, with one stop in Adel, Georgia (see Figure 12.1) to drop off one passenger.

VFR flights are flown in weather conditions where the pilot can fly and navigate without reliance on instruments within the plane. They require that the aircraft remain clear of all clouds and that visibility is three miles or better when flying on controlled airways and/or in the vicinity of major airports.

The weather earlier on the evening of June 6, 1995, over Georgia and Florida was characterized by nearly cloudless (clear) skies and good visibility (well above the three miles required for visual flight).

10:46 p.m.
The official weather observation made at the Gainesville, Florida, airport was transmitted on government communication lines to air traffic control facilities, Flight Service Stations, and other facilities concerned with aviation weather. These are made routinely every hour and whenever there is a significant change in the weather. This routine observation indicated weather conditions at the Gainesville, Florida, airport were:
- clear skies (no clouds);
- visibility (horizontally at ground level) of six miles in haze;
- temperature of 78°F;
- wind from 240° (southwest) at 8 mph;
- altimeter setting (a value the pilot could dial into the aircraft altimeter so it would correctly read heights at the field locations) of 28.95 inches (of mercury in a mercurial barometer).

These conditions were appropriate for a VFR flight at Gainesville, Florida.

11:00 p.m.
Pilot Ross DeWitt—a member of the aviation faculty at the University of Florida at Gainesvillle, Florida—entered the Kenn-

Air fixed base operation (FBO) center in Gainesville, Florida, a facility on the airport northeast of the main passenger terminal utilized by private aircraft, to prepare for a flight to Tallahassee, Florida. He conducted the preflight inspection of his aircraft and filed by telephone an instrument flight rules flight plan from Gainesville to Tallahassee, Florida, with the FAA Flight Service Station at the Gainesville airport.

The flight plan was filed IFR because of the reported low clouds in Tallahassee, which would require IFR flight toward the final portion of the trip.

The FBO was closing as he prepared to leave. He noticed a woman on the ramp waiting for her family to arrive from Covington and Adel, Georgia.

11:10 p.m.   The McNair flight left Adel, Georgia, en route to Gainesville, Florida.

11:14 p.m.   Ms. McNair contacted the FAA Air Route Traffic Control Center at Jackonsville, Florida (JAX ARTCC), by radio and indicated that the aircraft was seven miles south of Adel, Georgia, at an altitude of 2,500 feet (above sea level), climbing to 5,500 feet [VFR traffic flies at levels ending in 500 (i.e., 3,500, 4,500, and 5,500), while IFR traffic flies at levels ending in 000 (i.e., 5,000, 6,000, and 7,000)].

The ARTCC at Jacksonville assigned the McNair aircraft a transponder code (which automatically identifies the aircraft on the controller's radar scope at each radar hit every 12 seconds and provides the controller with the identity, altitude, and geographic location of the plane). The Jacksonville controller would continue to track the progress of the plane by means of his radar in a procedure known as flight following.

11:15 p.m.   As the McNair flight continued south-southeast toward Gainesville, a commercial airliner en route to Gainesville from Charlotte, North Carolina, was flying southwest toward Gainesville and descending from 23,000 feet. They had the lights of the

Gainesville airport in sight from 23,000 to 3,000 feet, at which altitude the crew lost sight of the field lights.

They requested a radar vector (direction and distance) to the field and approval for an instrument landing system approach (ILS)—a radio beam that can be followed by instruments within the plane to the runway.

While on that approach, the crew again saw the runway lights when they were within two to three miles east of the airport; they landed toward the west on the runway most nearly oriented to face into the wind coming from the southwest at 8 mph.

11:32 p.m.   The Jacksonville controller told Ms. McNair of the commercial flight's loss of sight of the airport while approaching and asked if she was equipped and qualified for instrument flight. She replied she was not. The controller informed her of the 10:46 p.m. Gainesville weather report, which indicated clear skies with six miles of visibility and light winds from the southwest (VFR weather).

11:42 p.m.   Ms. McNair advised the Jacksonville ARTCC that she was descending from 5,500 to 2,000 feet. The center controller acknowledged her report and advised her to maintain VFR, that is, not to enter clouds or reduced visibility (if encountered) that would require the use of instruments to fly.

11:43 p.m.   The controller in charge at the Gainesville Flight Service Station (GFSS), who was the official weather observer from 4:00 p.m. to midnight made an official weather observation and determined the sky was six-tenths covered by clouds, and the visibility was six miles in haze.

The six-tenths coverage of the sky by clouds constituted a "broken ceiling," the height of the base (bottom) of which was "measured" by a light beam to be 600 feet above ground.

The observation was recorded on the weather log (observational form) and entered into the computer, which displays such information at various personnel locations within the station.

The observer did not verbally announce this to any of the personnel at the facility, and none noticed it on their computer displays.

This official observation included a ceiling less than 1,000 feet above ground. The conditions required the field be "closed" to VFR traffic and required IFR flight for landings or takeoffs at the field, though none of the staff were aware of the observation and took no action nor was the observation relayed to the facilities such as the ARTCC at Jacksonville, as it should have been.

No one told Ms. McNair, who was attempting to locate the lights of the field, that the weather had changed and VFR conditions no longer prevailed, since no one—other than the observer going off duty—was aware of those critical facts.

11:50 p.m.    Pilot DeWitt received takeoff instructions and took off to the southwest. At about two miles from the field, while climbing through 500–600 feet above ground, he unexpectedly encountered the layer of clouds that had not been reported to him. He estimated these to be about 100 feet thick and found them somewhat disorienting, though he was qualified to fly IFR (on instruments).

He did not report this encounter to the Gainesville Flight Service Station.

11:51 p.m.    An employee at the Kenn-Air FBO called the GFSS and said a woman [the mother of twin four-year-old girls on the McNair plane] was waiting for an inbound private flight and asked if the GFSS had anybody en route.

The person at GFSS replied they were not aware of any inbound VFR traffic but would call JAX ARTCC and see if they had any inbounds to Gainesville.

11:52 p.m.　　The GFSS employee called JAX ARTCC and was told, "I got a Cherokee that's getting flight following…I do not know the point of departure hang on just a minute I'll call you right back" (Wentworth 1995, p. 3).

11:55 p.m.　　JAX ARTCC advised GFSS the inbound plane was being flown by Ms. McNair and was estimated to arrive in about 15–20 minutes.

11:58 p.m.　　JAX ARTCC asked Ms. McNair if she had Gainesville airport in sight. She replied, "Negative, I'm looking" (Wentworth 1995, p. 4).

The controller suggested a small heading change and said the airport would be seven miles straight ahead.

When the flight was three miles from the airport, Ms. McNair reported she could see the rotating beacon at the airport and no longer needed radar assistance.

### June 7, 1995

Midnight　　JAX ARTCC instructed Ms. McNair to change her transponder code and radio frequency and to call the GFSS on a new frequency "for advisories."

For the next two minutes Ms. McNair attempted to do so to no avail.

12:04 a.m.　　Ms. McNair, unable to contact GFSS by radio, switched back to JAX ARTCC and said, "I'm over the Gainesville airport. I cannot get anyone to pick me up there [to respond to my radio calls]. Do they operate at night?" (Wentworth 1995, p. 4).

She was advised that the tower was closed, but the Flight Service Station was open and used the frequency given her. The controller advised he'd call them on a land line, which he did.

The GFSS employee had left his radio switched to a headset, which he was not wearing, instead of to a loud speaker. When

JAX ARTCC advised him Ms. McNair was calling on the radio, he switched the radio to the speaker on his console.

12:05–
12:08 a.m.
For the next three minutes, Ms. McNair made two circles to the left over the Gainesville airport while talking to the GFSS person. At first she was unable to see the runway lights (which were dimly illuminated on three runways).

The GFSS man agreed to turn the lights on the east–west runway brighter and to dim the others. Ms. McNair then saw them, while directly overhead and thanked him for his help.

She then flew toward the east-northeast, presumably to line up with the runway and land heading west, after descending from 1,300 feet.

12:12–
12:18 a.m.
GFSS called Ms. McNair's plane three times and asked, "Are you clear of the runway?" (Wentworth, 1995, p.5) [presuming she had landed]. There was no response.

The GFSS in-flight specialist who had lost communications with the McNair plane went outside to attempt to see or hear the airplane but could not either see or hear one. He noticed what he believed to be "scattered clouds" (less than half the sky covered) with visibility between five or six miles but did not record an official observation.

12:29–
12:31 a.m.
The next official weather observation was made and recorded as "clear, visibility 5 miles in fog; winds from 250° at eight knots; altimeter 29.87" (Salottolo 1995, p. 2).

11:30 a.m.
Search and rescue personnel located the wreckage of the aircraft approximately five miles east-northeast of the Gainesville airport with all six occupants dead.

The relatives of the deceased occupants, as plaintiffs, alleged the crash was due to the negligence of Federal Aviation Administration employees in failing to properly monitor the weather and to correctly advise the pilot. They sought compensation from the U.S. government. The government alleged the crash was due solely to the negligence of the pilot, Ms. McNair, and denied that the weather conditions along the route or at the airport were unsuitable for VFR flight.

## THE TRIAL

The trial was held in the U.S. District Court for the Northern District of Florida, Gainesville Division, in the summer of 1998 before Senior District Judge Maurice M. Paul. All the above facts were made known to the court and became a part of the record either by stipulation by the attorneys, fact witness testimony, or excerpts of the "Factual Report" of the National Transportation Safety Board (NTSB 1995), which investigated the accident. Weather testimony was given by several fact witnesses, either as written statements or verbal testimony at trial. The defense attorney—for the government—engaged a forensic meteorologist they frequently used to testify as to their belief that the sky over the Gainesville airport was cloudless at midnight when Ms. McNair arrived to land there. He was a Certified Consulting Meteorologist with a Ph.D. in meteorology, who had never been an official weather observer or forecaster.

The plaintiff's attorney contacted Climatological Consulting Corporation and sought my services to assist the court in its search for the truth by testifying what the weather conditions were along the route of flight and at the Gainesville airport during the evening of June 6 and early morning of June 7, 1995 (VFR en route but IFR at the airport because of a 600-foot-aboveground broken cloud layer covering six-tenths of the sky). He had selected me partially because I had been a weather briefer and forecaster at Washington National Airport, that the subject of my thesis for my master's of science degree had been fog and stratus (cloud) formations over the southeastern United States, and that I had two years of experience as a meteorologist at the Naval Air Station in Jacksonville, Florida.

The fact witness statements and/or testimony included the following:

1.  Ms. Cox, the official weather observer on the 4:00 p.m. to midnight shift, in a statement to the NTSB investigators, which was included in "facts" at the trial, said she walked out the door of the Flight Service Station at about 11:43–11:44 p.m. (EDT) on June 6 and realized that clouds had moved into the area, so she went to the other side of the building to the ceilometer (a light beam and angular measuring device utilized to determine the height of cloud bases by geometric calculations) and got

"a good [light] spot on the [cloud] ceiling" measured to be at 600 feet "over the building and extending outward" (Wentworth 1995, p. 9). She estimated that the clouds covered six-tenths of the sky, constituting a ceiling of "broken clouds."

These conditions would not permit a VFR landing at the airport during the time they existed.

Her observed, but not fully distributed, 11:47 p.m. observation would indicate the official weather conditions at the airport until replaced by a later official observation, which was not made until 12:30 a.m., 20 minutes after the estimated time of the crash of the McNair airplane.

Ms. Cox left the airport at midnight and was aware of an aircraft going southeast at a fairly low altitude.

2. Pilot Ross DeWitt, who had filed his flight plan from Gainesville to Tallahassee shortly after 11:00 p.m., filed a statement with the NTSB on July 7—a month after the accident. It was entered into the trial record. It indicated that while he was on the ground near the FBO at about 11:00 p.m., he was not aware of the presence of clouds (which had not been mentioned in the weather briefing he received).

He was surprised to encounter a layer of variable scattered to broken clouds at about 500 feet immediately after take-off (at 11:50 p.m.).

3. Ms. Holly Saul (née Powell), who was the lady waiting at the FBO for the McNair flight to arrive, testified at trial that the conditions were so cloudy between 11:00 and 11:50 p.m. (EDT) that she could hear aircraft overhead but could not see them and that the sky condition "was patchy, with some [areas of] clear sky" (Paul 1998, p. 8).

4. Mr. Robert L. Gibson—the flight service specialist at Gainesville who had been in radio communication with Ms. McNair—went outside the GFSS building to see if he could physically see or hear her plane at 12:15 a.m. but neither saw nor heard a plane. He estimated the weather was "scattered clouds—with 5–6 miles visibility" (Wentworth 1995, p. 16).

5. The next official weather observation was made by FSS personnel between 12:29 and 12:31 a.m. on June 7 and recorded as clear; visibility 5 miles in fog; winds from 250° at 8 knots.

6. The government's expert weather witness testified that his opinion was that the sky was clear of clouds at the time Ms. McNair was over the airport at Gainesville and that she should have seen the ground lights outlining the runway and been able to make a VFR landing as far as the weather was concerned. He further testified that the 12:31 a.m. (EDT) weather observation (clear with five miles of visibility) was consistent

with the prevailing conditions at the time of Ms. McNair's arrival, and Ms. Cox's 11:43 p.m. observation of a broken cloud ceiling at 600 feet was "out of sync" with the preceding and following observation and at best was "but a temporary condition" (Paul 1998, p. 16).

7. My testimony on behalf of the plaintiffs was that Ms. Cox's official observation at 11:43 p.m. (EDT) on June 6 accurately represented the IFR conditions present at both the time of the observation and at the time of Ms. McNair's arrival 13 minutes later and was consistent with the experience of both the arriving commuter flight from Charlotte and the departing flight of Mr. DeWitt to Tallahassee.

8. Both the plaintiff and the defense attorneys put pilot performance experts on the stand. The government's expert opined, "Pilot McNair exercised poor judgment in continuing to fly to the northeast when at 0008:00 hours she was over the airport and had the runway lights in her sight [looking directly down she could see the runway lights], and this poor judgement on her part contributed to causing the accident" (Paul 1998, p. 17).

The plaintiff's pilot expert expressed the opinion that Ms. McNair flew northeast to descend and line up with the runway for a landing and became spatially disoriented when she unexpectedly entered a layer of broken clouds at 600 feet. Figure 12.2 illustrates how a pilot could see the runway lights from directly above the airport and lose sight of them when attempting to find them at a lower altitude with the existence of a 100-foot-thick layer of six-tenths of the sky covered by broken clouds at 600 feet above ground. When looking directly down at lights on the ground at least four-tenths of the downward view is unobstructed and the other six-tenths required viewing through a 100-foot layer of clouds (view A). When looking for the field while at a lower altitude, the viewer had no unobscured line of sight, and the slant range view through the clouds requires viewing through hundreds of feet of intervening cloud (views B and C).

The judge heard from economists produced by both plaintiffs and defendants as to the economic losses resulting from the crash. He ruled the crash resulted from negligent acts committed both by Ms. McNair and by government employees, and he ruled the faults were 60% by Ms. McNair and 40% by the government personnel. The court determined the total losses sustained by the survivors of the six persons on the aircraft was $4,569,445 of which the U.S. government was responsible for 40% because of its negligence, awarding the McNair plaintiffs $1,827,778 in damages.

# APL CHINA

*Christmas Overboard, North Pacific Ocean, 1998*

## THE EVENT

On October 26, 1998, the huge container ship, the *APL China*, carrying over 4,000 containers from Asia to North America was overtaken and overwhelmed by a raging North Pacific storm. The loss overboard of 388 containers and damage to 400 others was called the largest container casualty in history. Lawsuits totaling over $100 million were filed by the intended recipients of the lost or damaged cargo.

## THE SHIP

Container ships came into being shortly after World War II. They carry all their cargo in truck-sized containers. They form the present-day common means of commercial, long-distance, maritime freight transport. The first container ship (class C-1) was a converted surplus World War II tanker that carried 56 containers from Newark, New Jersey, to Houston, Texas, in April 1956. Soon thereafter, specially designed, ever larger container ships were built, each successive class carrying ever more containers in deep holds of the vessel and stacked on deck above the holds. Soon the larger classes of container ships were unable—because of their length or width—to pass through the locks of the Panama Canal. They became largely ocean specific.

Their capacity is measured in terms of 20-foot equivalent units (TEU), the number of standard 20-foot containers—measuring 20 feet long, 8 feet wide, and 8.5 feet tall—that a vessel can carry. Most ships today carry a mix of 20-foot-long and 40-foot-long containers (a 40-foot container is 2 TEUs) as well as special refrigerated containers of various sizes. One of the very largest modern container ships—the *Emma Maersk*, a C-13 class—carries close to 15,000 TEUs. The *APL China*, a class C-11 container ship, was built in Germany in 1994–95 and delivered to the APL shipping company in late 1995 to begin regular service between Southeast Asian and U.S. West Coast ports; 906 feet long, with a berth (beam width) of 131 feet and a draft (distance from sea surface to bottom of keel when fully loaded) of nearly 46 feet, the container carrying capacity was 4,832 TEUs plus 310 refrigerated containers. Fully loaded, the *APL China* carried approximately 3,500 containers in the deep holds of the ship and 1,355 above deck in rows of containers stacked as many as six (or more) high. The *APL China* carried a crew of 21 (no passengers) and was powered by an 11-cylinder, 66,385-horsepower diesel engine that propelled the ship at a speed of 24.6 knots (28.3 mph). It had a cruising range of 20,500 nautical miles.

## THE 30E VOYAGE

On October 20, 1998, the *APL China* departed Kaohsiung, Taiwan, and began the 10-day, nearly 6,000-mile voyage bound for Seattle. It was the *APL China*'s thirtieth eastbound (30E) voyage with products from Southeast Asia (principally China and Taiwan) destined for U.S. markets. The goods on this voyage were to be delivered to sales outlets in major U.S. cities—many on the East Coast—in time for the 1998 Christmas markets. During the voyage, the *APL China* captain routinely monitored radio broadcasts of weather forecasts for 48 to 96 hours in advance from the Japan Meteorological Agency (JMA) and the U.S. NWS. On October 24, when the *APL China* was four days into its voyage, an atmospheric low pressure center formed off the east coast of Asia, about 1,000 miles southwest of the position of the ship. Forecasters at the Japanese Weather Service in Tokyo, the U.S. National Weather Service near Washington, D.C., the U.S. Navy Fleet Numerical and Oceanographic Center in Monterey, and at the Ship Routing Company were in basic agreement that the storm would intensify and move rapidly northeastward toward the Gulf of Alaska.

As predicted, the storm intensified rapidly as it moved northeastward over the warm waters of the western Pacific. On October 25, the storm, moving faster than the ship, crossed the track of the *APL China*, well behind the ship, and on October 26, its center was located about 120 miles north of the ship (see Figure 13.1). Storm winds of 50–70 knots (60–80 mph) extended outward nearly 300 miles from the storm center, with wave heights of 30 to 43 feet. Caught in the most intense quadrant of the storm, with the strongest winds and highest waves, the *APL China*'s excessive rolling caused 388 containers to break loose and plunge into the Pacific Ocean, while 400 others were damaged but remained on board together with the relatively undamaged 3,500 below deck in the ship's holds. The storm, later named the *APL China* storm, has been classified by various analysts as an explosively intensifying low (EIL), a "Bergeron bomb," and more recently a hurricane force extratropical cyclone (HF cyclone).

An EIL is a low pressure center in which the central barometric pressure drops at a dramatic rate, such that the storm's intensity increases "explosively." A Bergeron bomb is defined as an extratropical surface cyclone in which the central pressure falls on the average of one millibar per hour for 24 hours. It is a maritime cold-season (October–April) event common to western portions of both the Pacific and Atlantic Oceans. A HF cyclone is a relatively new term introduced to the scientific literature in 2005 after a four-year study of extremely violent storms occurring over the North Pacific and North Atlantic Oceans in which hurricane force winds of over 64 knots (75 mph) occurred. Utilizing satellite reports of surface winds, analysts at the National Weather Service Ocean Prediction Center (OPC) studied 120 such storms in the three seasons between October 2001 and April 2004; there were approximately 20 such storms per year per ocean. The HF winds in all 120 occurred in the southeast quadrant (right-front quadrant) of the 120 storms, all of which were moving rapidly northeastward, as was the case with the *APL China* storm. The late October 1998 North Pacific storm, the *APL China* storm, was all of the above. Storms classified as Bergeron bombs are EILs, and, if they have winds of hurricane force [64 knots (75 mph)], they are HF cyclones.

## WAVES

Waves are generated by the winds blowing over the water surface. The stronger the winds and the longer they blow, the higher the waves become. The greater distance they blow over the water, the higher are the waves—measured from crest to trough—and the greater is the distance (wave length) between the crests. Most wave's heights, lengths, and periods (the length of time between the passing of successive waves) are observed from ships or coastal sites that do not have mechanical or electronic measuring devices to detect and record the wave characteristics. Thousands of estimates by experienced mariners have been compared with measurements made by such devices and their relationships have been analyzed. These comparisons show that the typical estimate of wave heights represents the average of the highest one-third of the wave heights measured by such devices. These estimates are called "the significant wave heights." The relationship of significant wave heights (as measured) to other measured waves is as follows:

| Wave | Relative height |
|---|---|
| Average | 0.6 |
| Significant | 1.0 |
| Highest 10% | 1.29 |
| Highest (one in a thousand) | 1.87 |

Thus, if significant wave heights are observed to be 20 feet, or 40 feet, these values would be as follows:

| | | |
|---|---|---|
| Average | 12 feet | 24 feet |
| Significant | 20 feet | 40 feet |
| Highest 10% | 25.8 feet | 51.6 feet |
| Highest | 37.4 feet | 74.8 feet |

The comparisons show that experienced mariners' estimates of wave length, period, and "direction of primary waves" (the direction they come from) are reasonably accurate.

Modern satellites now finely scan the oceans and sense the wind directions and speeds from their impact on the ocean surface. Computers routinely compute the wave heights, periods, and primary directions from known formulae relating winds to waves. Figure 13.2 shows a segment of the significant wave height and primary wave direction chart for 1200 UTC* on

---

* UTC time definition: "Weather observations around the world (including surface, radar, and other observations) are always taken with respect to a standard time. By convention, the world's weather communities use a twenty four hour clock, similar to 'military' time based on the 0° longitude meridian, also known as the Greenwich meridian. Prior to 1972, this time was called Greenwich Mean Time (GMT) but is now referred to as Coordinated Universal Time or Universal Time Coordinated (UTC). It is a coordinated time scale, maintained by the Bureau International des Poids et Mesures (BIPM). It is also known as 'Z time' or 'Zulu Time'" (National Hurricane Center 2015).

October 26, 1998, prepared by the U.S. Navy Fleet Numerical Weather Center, Monterey, California, with the position of the *APL China* added. It shows the primary waves coming from the southwest with significant wave heights of 36 feet at the location of the ship at that time. To the west-southwest of the *APL China*'s location, an area of waves coming from the west with significant wave heights of 48 feet is analyzed. This area (to the south of the storm center) was moving northeast at the storm speed of 30–35 knots and overtook the *APL China* later in the day. These wave heights were verified by two satellites in polar orbit specially designed to directly measure ocean wave heights. The sea state* entry on the *APL China* deck log at 1200 UTC on October 26 read "9" (on a scale of 9). This is described as "phenomenal" and indicates waves over 45 feet in height.

All ships are subject to a variety of forces because of the impact of the wind and waves on the ship and its cargo. The stacking of large numbers of containers above deck greatly increases the area for these forces to impact. The motions of the ship in response to these forces are described as pitching, rolling, heaving, surging, yawing, or swaying, often in combination of several of them. Pitching is a plunging and tossing motion with the bow and stern of the ship rising and falling (stern rising while bow falls and vice versa). Rolling is a swaying or side to side rotating motion in which one side drops down while the other rises. Heaving is a rising or falling of vertical motion of the entire ship. Surging is a change in the fore (or aft) motion of the ship— a sudden positive or negative acceleration of the entire ship. Yawing is an unintentional change right or left from the intended course, and swaying is a sideways movement of the entire ship. These are illustrated in Figure 13.3 (ship movement at sea), where the green up and down arrows at the bow and stern show the directions of pitching motion, the red curved arrows show the direction of rolling motion, the upward and downward small blue arrows show the direction of heaving, the blue arrows pointing forward at the bow and aft at the stern show the direction of surging, the curved purple arrows show the change in direction (right or left) from the intended course, and the large horizontal blue arrows show the lateral movement of the ship because of swaying. In fair weather with light winds and relatively calm seas (little or no wave action), these motions are small. They increase dramatically with higher winds and waves.

---

* "Sea state: A description of the properties of sea surface waves at a given time and place. This might be given in terms of the wave spectrum, or more simply in terms of the significant wave height and some measure of the wave period" (AMS 2012d).

At the height of the storm encountered by the *APL China*, the 906-foot vessel likely pitched as much as 5° or more (see Figure 13.4). At the pitching angle, the rise and fall of the bow and stern was nearly 100 feet in each pitching cycle. The foremost and rearmost container stacks underwent a series of dramatic elevator rides. Crew members testified that the roll angles of the ship reached 40° right and left at the worst of the storm. The container stacks therefore swept back and forth through an arc of 80°. When the bow was pitched down, it and the forward stacks of on-deck containers were submerged and waves broke against them. At the times of greatest roll, huge waves swept along the containers on the low side of the vessel adding greatly to the forces acting on the stacks of containers.

When extreme weather conditions are encountered, it was generally believed a vessel should be slowed and headed nearly directly "head-on" into the predominate waves. This is what the captain of the *APL China* attempted, unaware that a special phenomenon known as "parametric rolling" could affect his ship. Parametric rolling occurs when a combination of circumstances "excites" synchronous combinations of pitch and roll motions, resulting in the abrupt onset of very large roll angles of up to 40° port and starboard (see Figure 13.5). These circumstances include the following:

- very high waves in which the flared bow of the ship may periodically be submerged while pitching;
- very long waves with at least one or two ship lengths between wave crests; and
- the pitch and roll periods synchronized such that the pitch period is exactly twice that of the roll period.

When the *APL China* was slowed and headed nearly head-on into the predominate seas, the vessel was struck by other huge waves at an angle from the predominate waves, causing yaw angles of as much as 20° and pitch angles of as much as 10°, with the flared bow submerged at times and roll angles of 8°–12°, which were occasionally "excited" by parametric rolling to 30°–40°. During the approximately 12 hours—mostly at night—that the *APL China* was trapped in the most intense part of the storm, the primary waves increased in height from 30 feet with a period of 11 seconds to 43 feet with a period of 14 seconds, coming first from the southwest and later from the west. Aside from the primary waves, there were others (generated earlier in the storm) crossing these and creating "confused" and tumultuous wave conditions. During these 12 hours, various log entries on the *APL China* deck log included remarks of:

- gale force weather;
- inability to take deck rounds because of strong winds;
- requested help to secure loose chemical drums;
- vessel pitching and rolling heavily in confused seas and swell;
- altered course to 165° true to avoid head swells;
- cargo hold high water bilge alarm sounded;
- total power blackout (at 6:31 a.m., ship's time);
- power restored (at 7:45 a.m., ship's time);
- course and speed as per master's order because of stormy weather;
- steering west-southwest to run away from storm (at 6:23 a.m., ship's time); and
- vessel experienced violent storm and suffered extensive loss of containers on deck; suffered some damage on deck structures.

It was later learned that the "total power blackout" from 6:31 to 7:45 a.m. was caused by seawater entering a generator room high on the starboard side of the ship, which caused an electrical failure that shut down the computer controls to the main engine, causing it to shut down for over an hour, leaving the ship wallowing in the high seas. The captain later stated that it was as though the devil had taken control of his ship.

With these violent motions, a third of the above deck 1,355 containers were lost overboard and another third were badly damaged. The storm moved rapidly away toward the northeast. Winds and waves diminished and the *APL China* headed for Seattle, where it arrived late and with containers leaning over its side. The headlines in the November 23, 1998, issue of the *San Francisco Business Times* read: "Cargo goes overboard; Insurance lawyers surface" (Ginsberg 1998).

## LITIGATION

The intended recipients of the missing and damaged cargo failed to receive it for their Christmas sales. Many cargo interests filed suits against the shipping company alleging their losses (totaling over $100 million) were the result of negligence of the APL's employees and agents in the loading and securing of the cargo (containers) and the operation of the vessel in relation to the storm. The dozens of cargo loss damage claim suits were to be heard in Judge Charles Haught's Federal Court in New York. As is usually done, the many cases were consolidated to be heard as one.

A plaintiff steering committee was formed under the leadership of Attorney Raymond Hayden of the New York law firm Hill, Rivkins, and Hayden, which had a long and distinguished history of representing cargo interests in maritime disputes. Ray Hayden would become the lead attorney for all the plaintiffs in the court action. His credentials included having served as the president of the Marine Law Association of the United States. The defense legal team was headed by John Kimball of the New York–based Healy and Baillie law firm. He, too, had a long and distinguished history of defending maritime interests in litigation. Judge Haught, Ray Hayden, and John Kimball were well acquainted; each had long experience in maritime law and were widely recognized as extremely competent individuals.

In preparation for trial, depositions were taken of fact and expert witnesses in cargo loading, engineering, ship handling, and dozens of disciplines, including weather. Hill, Rivkins, and Hayden sought the services of Climatological Consulting Corporation, while Healy and Baillie engaged those of Dooley Seaweather Analysis, Inc., of City Island, New York, headed by Dr. Austin L. Dooley to provide expert weather testimony. The attorneys for the shipping company argued that the damages were the result of an act of God in creating an unprecedented, unforeseeable, and overwhelming storm against whose winds and waves they were helpless; that its intensity and movement were unprecedented and unpredictable; and that the master and crew were helplessly at the mercy of the overwhelming weather events and thus blameless. They sought to be absolved of all liability beyond the $55 million value of the vessel. The plaintiff's argument that the storm was expectable at that location and season required that there was demonstrable precedence of such an event. This required an analysis of the late October climatic history of weather events over that portion of the North Pacific Ocean traversed by the route of the *APL China* and a comparison of the weather encountered by the ship on its voyage 30E with the long-term weather history along that route.

Shortly after World War II, all maritime nations agreed to exchange their large archives of ship weather observations for the purpose of analyzing historical data to provide estimates of the likelihood of future expectable conditions in oceanic areas and seasons of the year. The U.S. repository of these international historical marine weather data was located in Asheville, North Carolina. It is today known as the National Climatic Data Center and serves both as a U.S. national data archive (and data processing facility) and as "World Data Center A" for the World Meteorological Organization. The U.S. Navy contracted the NCDC to prepare marine climatic atlases for each of the major oceans of the world and made these openly available to maritime interests (U.S. Office of Climatology 1959). Of the many charts and data tabulations, one set dealt with the climatological history of cyclonic storms month by month.

Figure 13.6 shows the area of cyclogenesis (cyclone storm formation and development) as well as the paths and speeds of such storms for the month of October over the western North Pacific Ocean. The planned and actual track of the *APL China* is overlain on the chart. The red hatched area on the chart is the principal area of cyclogenesis (development of cyclonic storms) historically observed in October over the North Pacific Ocean. The blue arrows show the historically preferred tracks of such developing storms in that month. The black lines labeled initial route and actual route showed the planned route of the *APL China*'s voyage 30E and the route actually traveled. After the ship passed through the red hatched area where October cyclonic storms historically were most likely to form, one did. The rapidly developing and intensifying storm moved along the historically "preferred track" as shown by the blue arrow and crossed the track of the ship, well behind it, but at a speed nearly 1.5 times the 24-knot (28 mph) speed of the ship. The climatological atlas data indicated that the storm of October 24–27, 1998, developed in the climatologically expected area of cyclogenesis in October and moved along the track of historical storms and thus had precedent.

The U.S. National Weather Service synopsis valid at 0600 UTC on October 25 and forecast valid at 1800 UTC October 26, 1998, are shown in Figure 13.7 as text (at the bottom) and "translated" pictorially with the synopsis in the lower left and the forecast in the upper right. The storm track and the track of the *APL China* are shown as black lines. Times of the storm and ship locations are shown as date/time notations such that 25/06 indicates October 25 at 0600 UTC, 26/18 indicates October 26 at 1800 UTC, and so on. In the synopsis at 25/06, the storm was located 450 nautical miles southwest of the ship with a central pressure of 985 millibars. Winds ahead of the storm were reported to be at speeds of 40 to 55 knots with "sea" (wave) heights of 14 to 28 feet. The storm was forecast to move to the location in the upper right of the figure in the next 36 hours (26/18) and to have a central pressure of 957 millibars, with winds of 55 to 75 knots in its southeast quadrant with seas of 28 to 42 feet. The projected track of the *APL China* would place the ship in that quadrant at that time, where the ship was overwhelmed by the severity of the storm whose winds and waves were "foreseen" by the forecasters and included in the forecasts routinely monitored via radio on the *APL China*.

The storm was both "precedented" and "foreseen," therefore not an "act of God." Both forensic meteorologists prepared—as required by the court in Federal Rule Of Civil Procedure 26—"expert witness statements," which provide the attorneys and the court with the details of the matters they expect to testify about at trial. An expert witness statement required that the witness provide advance written detailed information on:

A.  their identity;
B.  the subject matter of their anticipated testimony;
C.  the facts they relied upon;
D.  the opinions they have reached;
E.  the basis for their opinions;
F.  the data and information relied upon;
G.  exhibits they plan to show at trial;
H.  their recent publications;
I.  their compensation;
J.  their past expert testimony; and
K.  possible supplementation of opinions.

These were submitted, as required, well in advance of their depositions, providing the deposing attorney an opportunity to "discover" much of the information sought after in the discovery process and to prepare probing questions on all aspects of the witness' potential courtroom testimony. The probing of the witness during the deposition also provides the attorney an opportunity to observe the witness' demeanor and to assess the likely impact of their demeanor and testimony on the judge and jury at trial. A partial copy (with no figures) of my weather expert witness statement in the *APL China* litigation is included in the appendix at the back of this book. The key opinions in it dealt with the expectability and foreseeability of such a storm and their relation to the track of the *APL China*'s voyage 30E.

The plaintiff and defense weather experts were deposed by John Kimball and Ray Hayden sequentially, with other attorneys present and permitted to question the witness. Insofar as the basic facts were concerned, the witnesses were in basic agreement, differing principally on their opinions of the likelihood and predictability of such a storm being encountered. After discovery and the depositions were completed, the groups of opposing attorneys conferred. In public statements by both Ray Hayden and John Kimball, and by many trial lawyers, there is agreement that the majority of cases get settled before trial because pretrial discovery provides the attorneys and the court with good assessments of what the likely outcome of the trial would be. The more discovery provides information on which both plaintiff and defense attorneys can agree, the greater the chance of settlement. With the aid of Judge Haught, the plaintiff and the defense attorneys negotiated a multimillion dollar settlement. The extensive discovery and intensive depositions in this case led all involved to agree that settlement (without extended trial) was appropriate, and they negotiated an equitable apportionment of damage payments to all the claimants.

# CHAPTER FOURTEEN

# STATE OF NORTH CAROLINA V. MICHAEL PETERSON

*Cool by the Pool, Durham, North Carolina, 2001*

District Attorney James Hardin was preparing his case against Michael Peterson in early 2003. Michael Peterson had been indicted by a grand jury for the December 9, 2001, murder of his wife Kathleen in their million dollar home in Durham, North Carolina. Michael claimed Kathleen had fallen down the back stairs, but the police believed she had been brutally murdered. Michael said they had been sitting by the large, outdoor swimming pool sipping wine and discussing Christmas plans when Kathleen went inside to answer the phone at 11:30 p.m. that evening. He said he had remained by the pool smoking his pipe, sipping wine, and listening to the gurgle of the water circulating through the filters and the pool. Clad only in shorts and a T-shirt, he told the first responders, and later the grand jury, that he had sat there quietly until 2:30 a.m. when he went into the house and found Kathleen in a pool of blood at the foot of the stairs, still breathing. He called 911 at 2:40 a.m. The first responders were suspicious and called the police.

The first responders had recorded the outdoor temperature at 51°F. Though the weather was a minor factor in the case, James Hardin believed it was an item worth the jury's consideration in considering the truth of Michael Peterson's testimony. Mr. Hardin went to the National Weather Service at the Raleigh–Durham International Airport, nine miles from the Peterson house. He met with Stephen Harned, the meteorologist in charge.

Mr. Harned provided the district attorney with certified copies of the hourly observations taken at the airport on December 8 and 9, 2001, the night of Kathleen's death. Though these records were certified to be accurate copies of the original forms and were bound with a gold seal and a blue ribbon, making them admissible in court, Mr. Hardin felt live testimony from a professional meteorologist would be needed when they were presented to the jury. He asked Mr. Harned if he could appear at the trial to answer questions about the data and whether they represented conditions at the Peterson home. Mr. Harned was forbidden by government regulations to leave his duty station to testify in court but explained there were private sector board-certified consulting meteorologists who could do so. He provided the district attorney with a list published by the American Meteorological Society of Certified Consulting Meteorologists. Climatological Consulting Corporation was on that list.

Mr. Hardin had David Saacks, one of his assistant district attorneys, call me in late February 2003 to see if I could review the data and appear at trial, which was set for May 12, 2003, in the Superior Court for Durham, North Carolina. I agreed to do so. Soon, thereafter, I met with James Hardin, David Saacks, and Frieda Black (also an assistant district attorney) at their office in Durham. They discussed the pending trial. They did not believe Michael Peterson's persistent story about a pleasant evening by the pool discussing Christmas plans; Michael's stay by the pool from 11:30 p.m. to 2:30 a.m. clad only in shorts and a T-shirt while the temperature dropped slowly from 55° to 51°F was unlikely. The district attorney wanted me to:

- conduct a detailed analysis of the weather;
- relate the temperatures at the airport to those at the pool;
- relate the poolside temperature to any published articles related to human comfort;
- prepare illustrative charts and visuals for the jury to view; and
- be prepared to testify at the trial in Durham sometime in May or June.

He was aware that Michael Peterson's defense attorney would argue that the temperature at poolside could be very different from those at the airport and felt it important that an independent scientific reconstruction of these was essential.

The weather data—collected hourly at observing sites—were archived at the National Climatic Data Center and available to the public at a cost for reproduction. Utilizing data from the nearest six-hourly weather observing stations, maps of the surface temperature patterns over Virginia and North and South Carolina were constructed for several hours prior to and after the time Michael Peterson said he sat by the pool. The data from 10:00 p.m. on December 8 to 2:00 a.m. on December 9 are shown in Table 14.1.

**TABLE 14.1.** Temperatures °F (see Figure 14.1 for locations).

|  | December 8 | | December 9 | | |
|  | 10:00 p.m. | 11:00 p.m. | Midnight | 1:00 a.m. | 2:00 a.m. |
|---|---|---|---|---|---|
| Danville, Virginia | 51.08 | 51.08 | 51.08 | 51.08 | 51.98 |
| Greensboro, North Carolina | 50.0 | 51.08 | 51.98 | 53.06 | 53.96 |
| Raleigh/Durham, North Carolina | 57.02 | 55.02 | 53.96 | 51.98 | 51.08 |
| Rocky Mount, North Carolina | 62.96 | 62.96 | 62.06 | 60.98 | 60.08 |
| Goldsboro, North Carolina | 64.4 | 64.4 | 66.2 | 66.2 | 64.4 |
| Fayetteville, North Carolina | 64.94 | 62.96 | 62.96 | 60.98 | 62.06 |

Graphs of the hourly temperatures at the six stations from 10:00 p.m. to 2:00 a.m. were constructed. They showed the following:

- a slight rise in temperature at Danville (DAN) (47 miles northwest of the pool area) of nine-tenths of a degree (from 51.04° to 51.98°);
- a slow rise in temperature at Greensboro (GSO) (57 miles west) of nearly 4° (from 50.0° to 53.96°);
- a rather steady fall of temperature at Raleigh–Durham International Airport (RDU) (10 miles to the southeast) of nearly 6° (from 57.02° to 51.08°);
- a smaller fall of temperature at Rocky Mount (RWI) (58 miles to the east-southeast) of nearly 3° (from 62.96° to 60.08°);
- a small rise and then fall in temperature at Goldsboro (GSB) (30 miles southeast, from 64.4° to 66.2° to 64.4°); and
- a gradual fall and then slow rise of temperature at Fayetteville (FAY) (67 miles south, from 64.94° to 60.99° to 62.06°).

To see what these data truly meant in terms of space and time, and more accurately determine the sequential conditions at the house and pool in Durham, maps were made of the distribution of temperatures over the 114 mile (north to south) by 127 mile (west to east) area with the pool at its center for each hour from 10:00 p.m. to 2:00 a.m. December 8–9, 2001. These maps showed a mass of cold air—with the lowest temperature slightly below

50°—moved eastward at about 15 mph for 60 miles in four hours, from north of Greensboro to north of Raleigh–Durham. This movement of cold air followed the passage of a southeastward-moving cold front, which crossed the area late in the day on December 8. The temperature at the pool at each hour was interpolated.* Figure 14.2 shows the progression of these interpolated temperature values at the pool in relation to the time Mr. Peterson allegedly sat there, shown as the vertical yellow band on the graph, and the "comfort zone" of temperature (described below), shown as the horizontal green band, as well as the temperatures at the airport. The graphs show that the computed poolside temperatures dropping slowly from 55°F at 11:00 p.m. to 51°F at 2:00 a.m. were identical to those at the airport and similar to the 51°F recorded by the first responders on their arrival to the house.

The AMS Glossary defines the comfort zone as "the ranges of temperature, humidity and air movement under which most persons enjoy mental well-being" (AMS 2012e). These vary somewhat by season and the native climate of the persons involved. For the eastern United States at the latitude of Durham, North Carolina, in the winter months with nearly calm winds and humidity similar to that of the nighttime hours of 11:00 p.m. to 2:00 a.m. on the night of Kathleen Peterson's death, the comfort zone lies between 65° and 74°F. The temperatures at the pool side were 12° to 21° colder than the comfort zone that fateful night.

The case turned into a very high profile case to be broadcast on Court TV, lavishly reported in local and regional newspapers, and later the subject of several books and TV documentaries. After preparing the potential weather testimony, I met with Mr. Hardin and his two assistants at his office in Durham. They approved the preparatory work and agreed to have Frieda Black interrogate me during direct testimony at trial. I was called to testify in early July 2003. I was sequestered in a small room near the courtroom (not permitted to appear in the courtroom until called to take the stand). Late in the morning, I was called to appear, took the stand, was sworn in (to tell the truth, the whole truth, and nothing but the truth) by the clerk of court, and was presented by Frieda Black as an expert on weather.

The defense attorneys immediately requested to examine my credentials and experience in voir dire out of the presence of the jury. Their request was granted by Superior Court Judge Henry Orlando Hudson, and the jury was led out of the courtroom. The lead defense attorney, David Rudolf, launched

---

* Interpolation is the estimation of unknown intermediate values from known analyzed values (e.g., temperature).

a vigorous attack questioning my credentials, my familiarity with the Peterson home, my knowledge of the scientific literature on human comfort, whether I had taken any of the weather observations I relied on, and how I could testify about the weather in Durham when I was not there that fateful night. He then asked for two weeks to examine all my documents, the data, my written reports, and all my planned exhibits. Judge Hudson granted the request and I was escorted from the courtroom before the jury returned to wonder why I was gone.

The two weeks stretched into four before Mr. Hardin had the opportunity to call his weather witness back to the stand. I was sworn in again by the clerk of the court, queried again in the presence of the jury about my education experience and qualifications, and was offered as an expert on meteorology and climatology. Once again the defense attorney launched a blistering attack ending with a motion I be disqualified by the court. Judge Hudson denied the motion and directed Frieda Black to proceed with direct examination, which went well. The jury appeared genuinely interested. In answer to Ms. Black's final question, I responded that it was my opinion, based on the data, my analysis, and the scientific literature, that it was too cool by the pool for an individual clad only in shorts and a T-shirt to remain comfortably seated outdoors for nearly 3 hours without becoming uncomfortably chilled. In belligerent and vigorous cross-examination, the defense attorney attempted to cast doubt on the data, the literature, and all my testimony. They moved all my testimony be stricken from the record and the jury be instructed to disregard it. The judge denied their motion.

While the weather testimony was incidental in regard to the principal evidence, the telling of the personal histories of all those involved, the financial traumas, and the sometimes startling facts placed in testimony by all the witnesses, it was a factor in assessing Michael Peterson's credibility. He was found guilty of murdering Kathleen and ordered to serve a life sentence in prison. My involvement in this case ended here, though his complex story continued.

# CHAPTER FIFTEEN

# LOOKING BACK OVER THE YEARS

I grew up sailing, served in the navy, and was an oceanic weather forecaster. I had expected marine meteorology to be the cornerstone of my courtroom work, but aviation became predominant until late in my consulting experience. The aviation attorneys that engaged me became frequent clients and spread the word. Approximately 75% of my over 200 cases involved aviation weather. The data required are somewhat different for aviation and marine cases, but were all available from the National Climatic Data Center. Having served as the director helped me order the pertinent data with support and help from my old staff.

With these data I began my detective work. Forensic meteorology requires careful and accurate retrospective weather reconstructions. These often required very detailed and small time/space scale analyses—frequently relying on supplemental and/or non-standard meteorological data such as eyewitness statements, photos, police reports, NTSB reports, site visits, and so on.

When I started in this field, there might have been up to 50 other forensic meteorologists. Few were using visuals to show the details of their analysis and to illustrate the weather to the courtroom. I quickly learned the importance of these images and spent a lot of time trying to improve them. I greatly appreciated the help of my wife, Martina, an accomplished artist, who designed and perfected many of them. We started with hand drawn poster boards and evolved to large, commercially produced color images. From there, I graduated to carrying an overhead vugraph projector and screen to display transparencies to the court. I became known as the man with the pictures. My expensive analysis projector in the 1980s, which could project a time-lapse sequence of images, was a great improvement in showing sequential radar images. First used in the Alicia Tank 089 case, the jury was convinced of my testimony after seeing the rainbands moving over the stationary image of Tank 089 on the screen. With time and technology, televisions were installed in courtrooms, strategically placed before the jury, judge, and counsel. Computers and computer simulations have now taken over. While the technology in the courtroom was improving, the technology in the meteorological field dramatically advanced.

I saw NCDC continue to modernize and improve the quality and availability of the data. Many observing stations became automated and directly relayed the data to NCDC. Other improvements were radar and satellite observing equipment. New radar technology evolved with more station locations, offering both quality and quantity. Satellites improved as well from simple polar orbiting to advanced high-tech polar orbiters and geostationary satellites. Great improvements in the resolution of the images and availability of the data all helped in my analyses of the weather. CCC's associates, all of whom were CCMs, kept me abreast of these improvements in their specialty fields. With a dozen or so of them spread out over the United States, they all greatly contributed to CCC's success. They enjoyed having the company handle the business end of their activities, as well as the annual week-long retreats held in Asheville, North Carolina, where new technologies and ideas were shared. Other associate meetings were held during the AMS and National Council of Industrial Meteorologists' annual meetings.

I was very privileged to work with Dr. Fujita. Over the many years of our association, I continued to be impressed by his mental acuity. I have never worked with anybody who was so detailed and saw things everyone else overlooked. It was a great honor to work beside such a talented, perceptive, and kind individual, who gave so much of himself to his research.

After I moved to Asheville on April 1, 1961, I purchased a hilly 100-acre farm east of Asheville. It included a creek, pasture lands, pine forests and a variety of old buildings. In 1975, my wife and I built our dream house and I started my new company in the den with one secretary and a graphic studio area in the large basement. We outgrew this space within a year, and moved to an old one-room log cabin near the house. I increased the staff to include an assistant meteorologist who was a CCM. After two years, the staff and files grew in size and we moved to a remodeled house near the creek. This location, overlooking the large pastures, was a beautiful setting for the next 20 years. When I became frustrated by a technical problem, I would announce to the staff "I'm going out to mow," and would climb on my riding mower. Nature's panorama of blue skies, white clouds, hillside forests and green grass helped clear my mind and was conducive to constructive thinking allowing me to return to the office with a fresh view and a mowed pasture. The five-minute walk from home to the office made it very convenient for me when I was not travelling. And I did travel extensively to meet clients, visit accident sites, testify in court, and attend depositions and meetings. I became a 1.6 million miler with Delta Airlines.

I have been very fortunate to have had so many wonderful experiences, and to have known at age 4 that I wanted to be a weather man. My parents, mentors, and teachers' support of my strange fascination for the weather nourished my passion. My wife supported and encouraged my new adventure and served as vice president to CCC. Her input and artistic help were crucial to its success. Being a forensic meteorologist was exciting, challenging and filled with the fun of meteorological detective work. Writing this book has brought back fond memories of these great 26 years.

I hope you have enjoyed it.

# ACKNOWLEDGMENTS

I am extremely grateful to my parents who supported my fascination for the weather; Martina for her devoted support; my professors, bosses, and colleagues, who prepared me for my life adventures; all those people who made my third career delightful and challenging, including the clients, my staff, associates, and advisors without whom none of this would have been possible; Sarah Jane Shangraw, AMS Books Managing Editor, for giving me a chance to publish my first book; Beth Dayton, AMS Books Production Manager, for her and her staff's edits and help; my niece, Sylvia Forward, for her years of devoted help with this book; and my son, Bob (an attorney), for help in editing. It was such a pleasure to work with so many talented individuals.

# EXPERT WITNESS STATEMENT

A. **IDENTITY:** William H. Haggard, CCM curriculum vitae

B. **SUBJECT MATTER:** Reconstruction of the weather, waves, climatic conditions, and forecasts, related to the transpacific voyage of the M/V *APL China* on voyage 30E from Kaohsiung, China, to Seattle, Washington, October 20 to November 6, 1998, including the following:
   - climatological conditions over the North Pacific in October and November from historical marine data;
   - the weather events of October 24–28, 1998, in relation to the location of the *APL China* on those dates;
   - the weather experienced by the *APL China* October 24–27, 1998;
   - the analyses and forecasts of North Pacific weather made by the Japan Meteorological Agency (JMA), the U.S. National Weather Service (NWS), and Weathernews, Inc. (NWI-Oceanroutes); and
   - winds and waves experienced by the M/V *APL China* and their expectancy.

C. **FACTS:**
   1. The M/V *APL China* began an eastbound voyage (30E) of the North Pacific at Kaohsiung, China, loaded with containerized cargo, on October 20, 1998, bound for Seattle, Washington.

2.  The planned route of the vessel was from Taiwan, north of Hachijō-jima; then a great circle to 47°N, 175°E; then a rhumb line course to 50°N, 165°E; and then a great circle course to Cape Flattery (Mapped in Fig. A1, not shown here).

3.  Weather routing information was provided to the *APL China* by WNI-Oceanroutes.

4.  WNI-Oceanroutes recommended a more southerly route on October 24.

5.  The vessel was sailed along the recommended route.

6.  The vessel deviated to the right from the planned course, commencing on October 24, and took a series of courses as was shown in Fig. A1 (planned and actual routes of M/V *APL China* on voyage 30E, October 1998).

7.  A rapidly moving and intensifying atmospheric low pressure center overtook the M/V *APL China* and passed it on October 26, 1998, placing the vessel within the area of maximum winds and waves of the storm on October 26, 1998.

8.  The vessel was maneuvered between 0900 and 2000 UTC on October 26 (2200 indicates the first October 26, 1998, ship time and 0900 indicates the second October 26, 1998, ship time, after crossing the International Date Line) in a generally southerly direction with "courses and speed as per Master's order due to stormy weather" (per ship's deck log).

9.  The vessel lost power and was unable to maintain course for several hours on October 26, 1998.

10. There was an "extensive loss of containers on deck," damage to several remaining containers, and some damage to deck structures (per the deck log).

11. The storm center passed about 120 nautical miles north of the M/V *APL China* at about 1800 UTC on October 26, 1998.

12. The storm at that time was moving east-northeastward (gradually turning to northeastward) at about 30 knots.

13. The maximum winds and waves of the rapidly moving storm system were south of the low pressure center.

14. The maximum winds and waves of the storm system passed over the positions of the M/V *APL China* between 0900 and 2000 UTC on October 26, 1998 (2200 indicates the first October 26, 1998, ship time and 0900 indicates the second October 26, 1998, ship time).

15. After the storm moved rapidly northeastward past the M/V *APL China*, the weather moderated, and the vessel was able to return to a direct route towards Cape Flattery.

16. The weather over the western portion of the North Pacific Ocean was being analyzed on an ongoing operational basis by several organizations in October 1998, including the JMA, the U.S. NWS, the U.S. Navy Fleet Numerical Meteorological and Oceanographic Center (FLENUMMETOC), and Weathernews Incorporated (WNI).

17. The synopses (and forecasts) of the JMA are documented (and illustrated) together with segments of the track of the M/V *APL China* in Tab 1 accompanying this report.*

18. The synopses (and forecasts) of the NWS are documented (and illustrated) together with segments of the track of the M/V *APL China* in Tab 2 accompanying this report.

19. The analyses made by WNI are documented in Tab 3 accompanying this report.

20. The successive positions of the center of low pressure of a developing storm—as analyzed by these three organizations—are shown for October 24 through 28 in contrasting colors in Fig. A2 (not shown here).

21. The lowest pressure values at each analyzed time—as analyzed by each of these three organizations—are shown in Fig. A3 (not shown here), with the colors corresponding to those of Fig. A2.

22. While the precise low center locations and the analyzed central pressure values at each analyzed time varied somewhat from agency to agency, the locations, movement, and intensity of the storm center showed substantial agreement in the three analytical sequences.

23. All three analyses show a developing low pressure center near 36°N, 150°E at 0000 UTC on October 25 (1200 October 25, 1998, ship time), with central pressures approximating 990 millibars (hectopascals).

24. By 1200 UTC on October 26, 1998 (0100 indicates second October 26, 1998, ship time; 30 hours later), all three analyses showed the low center near 43°–44°N, 175°E with the central pressure approximately 960 millibars.

---

* The report was originally submitted in a binder with ten tabs containing the various data relied upon.

25. Then 20 hours later, by 1200 UTC on October 27, the three analyses showed the storm center between 45° and 50°N and east of 175°W, with the central pressure above 970 millibars.

26. Figures A4 and A5 (not shown here) show these analyzed storm positions and central pressures in relation to the track of the M/V *APL China*.

27. Climatological data and analyses have been prepared by various agencies to provide mariners with atlas information on frequencies of occurrences of various weather factors, storm tracks, areas of cyclogenesis, direction, and speed of movement of low centers, wave height expectancy, and other weather data pertinent to travel at sea.

28. One widely utilized atlas for the North Pacific Ocean is the *U.S. Navy Marine Climatic Atlas of the World, Volume II, North Pacific Ocean*, published by the U.S. Navy.

29. Charts of significant value for anticipating conditions on Pacific Ocean voyages are included for each month of the year.

30. Many of the climatological charts are followed by statistical tables or statistical graphs designated by numbers 1 through 45, which illustrate the statistical distributions of the element on the accompanying chart (e.g., the October surface winds chart is followed by a graphical and numerical tabulation of all the surface wind data from the 120 years of data utilized in the atlas within each of the 45 selected representative areas).

31. The two statistical areas closest to the place where the storm center overtook and crossed the track of M/V *APL China* on October 26, 1998, are areas 11 (north of the track) and 23 (south of the track).

32. Where appropriate, these statistical representations of the data in these two areas are contained in Tab 4, immediately following the chart of the weather element of concern.

33. Area 11, following the October surface winds chart in Tab 4, shows the most frequently observed wind direction in October in that sampling area is from the west, occurring 30% of the time; the second most frequent is from the northwest, occurring 18% of the time; and winds from the southwest are the third most commonly observed, occurring 14% of the time.

34. The numbers on the graph indicate the percentage frequency of wind occurrences by direction and speed. They show winds of 48 knots or greater occur in 2% of all October observations in that sampling area, with half of those (1%) occurring with west winds.

35. Similar statistics are shown for area 23 (nearest to the location of the encounter of the M/V *APL China* with the low pressure storm on October 26, 1998). They show that winds from the south, southwest, and west each have expectancies of 10% or greater, and that the expectancy of high wind speeds is greatest from winds from the southwest.

36. Linear interpolation between sampling areas is inappropriate if major storm tracks pass between the sampling areas.

37. The November surface wind data for these sampling areas are quite similar to those for October—as shown in the graphical/tabular material for areas 11 and 23 immediately behind the November surface wind in Tab 4.

38. Tab 4 contains climatological charts and statistical data for October and November surface winds, sea level pressure, mean wind, low wave heights, medium wave heights, high wave heights, low pressure centers (including areas of cyclogenesis), primary storm tracks, direction and speed of low centers, and tropical cyclones.

39. A late October voyage would be planned best by relying on climatic data for October and November, as the statistics change between midmonthly values. Late October expectancies would be between October and November values.

40. On each chart, a plastic overlay has been added showing the initial and actual routes of the M/V *APL China* on voyage 30E in October–November 1998.

41. The chart concerning "low pressure centers" for October, which includes cyclogenetic areas and speed and movement of lows, is reproduced as Fig. A6 (not shown here), with the initial and actual track of the M/V *APL China* on a clear, plastic overlay.

42. The blue arrow on the climatological chart of October low pressure centers shows the preferred primary storm track. It crosses the M/V *APL China* actual track at approximately 40°N, 160°E.

43. The speed and direction "roses" for the October movement of low pressure centers surrounding this point indicate a climatic history of October low pressure center movement toward the northeast at 27 to 31 knots as the predominate direction and speed.

44. In November, this predominate movement in this location is toward the northeast at 28 to 37 knots. Interpolation of the October and November values would indicate a late October expected storm movement in this area would be toward the northeast at 28 to 34 knots.

45. The red hatched area in Fig. A6 and the corresponding November chart in Tab 4 indicate the "principal areas of cyclogenesis."

46. The rapidly developing low pressure center of October 24–27, 1998, had its origin in the October/November climatologically expected "principal area of cyclogenesis" and moved in the direction of the October "preferred storm track" at a speed reaching 35 knots (a climatologically expectable speed in late October).

47. Meteorologists have defined an extratropical surface cyclone whose central pressure fall averages at least 1 millibar per hour for 24 hours as an "explosively intensifying low" (EIL), often designated a "bomb," so named by the Norwegian meteorologist Tor Bergeron.

48. "Bergeron bombs" (EILs) are predominately maritime, cold season (September–April) events.

49. These events are particularly common over the cooler ocean waters of the western Atlantic north of the Gulf Stream and the correspondingly cooler waters of the western North Pacific north of the Kuroshio Current.

50. Such "bombs" are classified in terms of "bergerons," where "1 bergeron bomb" has a 24-hour central pressure fall of 24 millibars; "2 bergeron bombs" have a central pressure fall of 48 millibars in 24 hours, and so on.

51. Some storms, such as the Queen Elizabeth II (QEII) Storm of September 1978 in the Atlantic and a storm of February 4–5, 1974, in the Pacific have reached 2.8–3.3 bergerons, with deepenings exceeding 50 millibars in 24 hours.

52. As many as 53 "bombs" per year, with 2% of them being "2 bergeron" or greater in strength, were found to have occurred over the Pacific (mostly western Pacific) in a study by MIT meteorologist Frederick Sanders and John Guakum made in 1980.*

53. The storm of October 26, 1998, affecting the M/V *APL China* had a pressure fall of 26 millibars in the 24 hours just prior to its lowest central pressure (and its impact on the vessel), thereby qualifying as a minimal "bomb."

54. Climatological data review prior to a trans-Pacific voyage would normally include the risk of encountering a tropical storm.

55. Tropical cyclone characteristics over the North Pacific are shown in the U.S. Navy atlas excerpts in Tab 4, with overlays of the M/V

---

* See Sanders and Gyakum (1980).

*APL China* voyage 30E routes added to the October and November charts.

56. They show a preferred western Pacific tropical cyclone route crossing the M/V *APL China* track during the early portion of the voyage in both months.

57. Wind is the motion of a mass of air having a common direction of horizontal motion with reference to the surface of Earth. It has both a speed and a direction from which it comes.

58. The speed and direction of wind relative to Earth's surface is determined by the resultant of several forces, the principal one of which is the pressure gradient force that is generated by the horizontal change of barometric pressure on Earth's surface, as measured perpendicular to the isobars.

59. On a moving vessel, the **apparent wind** (sensed on the vessel) is created by a combination of the motion of the vessel relative to Earth's surface and the motion of the air (**true wind**) relative to Earth's surface.

60. The apparent wind can be converted to the true wind by solving the vector triangle of apparent wind–wind relative to the vessel due to the vessel's motion = the true wind.

61. On vessels equipped with an anemometer (an instrument to measure the apparent wind), the calculations are made routinely from the indicated wind from the anemometer and a knowledge of the vessel's course and speed.

62. The **true wind** values are entered in the log as a part of the regular weather observations and (on a "weather reporting ship," such as the M/V *APL China*) transmitted via radio to agencies making weather analyses and/or forecasts; they are subsequently sent to data archives.

63. On ships without operating anemometers, wind direction and speed are estimated by the effect of the wind on the water surface. Direction is observed by viewing the wind streaks (if present) on the water surface and the speed estimated by the "sea state" generated by the wind.

64. Though the M/V *APL China* was equipped with an anemometer, it was inoperable during voyage 30E in October 1998, and wind observations had to be made visually from the sea state.

65. Admiral Sir Francis Beaufort devised a numerical scale for indicating wind speed in 1806. It was adopted in 1838 and has been extended since.

66. The Beaufort wind scale, with its speed conversions, descriptions, and effect at sea is shown in Table A1 (not shown here; after Kotsch's* *Weather for the Mariner*).

67. The wind data recorded in the log of the M/V *APL China* for the time period from October 24 to 27, 1998, are listed in Table A2 (not shown here), below, together with the sea condition data (which are discussed later in this report).

68. Log entries were made "in real time" on October 24, 25, and 27, but data for much of October 26 were not entered contemporaneously on that date. They were entered *post facto* on the 27th in the log for the 26th.

69. A copy of the October 1998 M/V *APL China* deck log book is contained in Tab 5 at the end of this report, followed by a time conversion table.

70. A plot of the time sequence of weather data from the M/V *APL China* deck log for October 24–27, 1998, is contained in Fig. A7 (not shown here). Wind direction is shown by the arrow shaft with the barbed tail at the end in the direction from which the wind is coming. The barbs (or feathers) in this figure represent wind speed with each 0.5 barb representing one Beaufort force such that 2.5 barbs indicates Beaufort 5; a "flag" (▲) represents force 10, and so on. Temperature in degrees Celsius is plotted in the upper left of each plot, dewpoint in degrees Celsius is plotted in the lower left, and barometric pressure in millibars is plotted in the upper right.

71. The wind (as entered in the ship's log) progressed from northeast force 8 at 0100 UTC on October 24 (1200 October 24, 1998, ship time) to southeast force 8 at 0200 UTC on October 25 (1400 October 25, 1998, ship time) to southeast by east force 10 at 0200 on October 26 (1500, first October 26, 1998, ship time) to south-southwest force 11 at 1300 UTC on October 26 (0200, second October 26, 1998, ship time) to southwest 10 at 1700 UTC on October 26 (0600, second October 26, 1998, ship time), becoming west force 10 at 1800 UTC (0700, second October 26, 1998, ship time) and west northwest force 10 by 2300 UTC (1200, second October 26, 1998, ship time) on October 26, 1998. By 1800 UTC on the 27th (0700, October 27, 1998, ship time), the recorded winds had diminished to west-northwest force 5.

---

* See Kotsch (1977, pp. 220–221).

72. Force 9 winds were recorded from 0800 UTC on October 25 (2000 October 25, 1998, ship time) to 1700 UTC on that date (0500, first October 26, 1998, ship time). Force 10 winds were recorded from 1800 UTC on October 25 (0600, first October 26, 1998, ship time) to 1200 UTC on October 26 (0100, second October 26, 1998, ship time). Force 11 then prevailed for four hours followed by force 10 for the next seven hours. Then force 9 prevailed for eight hours, followed by steadily diminishing winds.

73. At no time were winds greater than force 11 recorded on voyage 30E of the M/V *APL China*.

74. The barometric pressure (from the ship's log) is entered hourly at each wind entry by three or four digits, representing the pressure in millibars (e.g., the 1200 UTC entries are shown as October 24 = 1025 mb, October 25 = 1010 mb, October 26 = 968 mb, and October 27 = 1014 mb).

75. The lowest barometric pressure entry in the log was 968 millibars from 0900 UTC on October 26, 1998 (2200, first October 26, 1998, ship time), to 1200 UTC on that date (0100, second October 26, 1998, ship time).

76. Figure A8 (not shown here) shows the trace from the barograph from the M/V *APL China* from 2300 UTC on October 24 (1100 October 25, 1998, ship time) to 2100 UTC on October 28 (1100 October 28, 1998, ship time).

77. As the developing low overtook the M/V *APL China*, barometric pressure on the ship dropped steadily from 1020 millibars at 0000 UTC on October 25 (1200 October 25, 1998, ship time) to 968 millibars at 0900 UTC on October 26 (2200, first October 26, 1998, ship time). It remained at 968 millibars for four hours while the ship was at its closest point southeast and then south of the passing low center. The pressure rose rapidly during the remainder of October 26 as the low center moved rapidly northeastward away from the ship. It reached 1006 millibars by 0000 UTC on October 27 (1300, second October 26, 1998, ship time).

78. The large variations in the barometric pressure trace on the barogram from 0700 UTC on October 26, 1998 (2000, first October 26, 1998, ship time), to 2000 UTC on that date (0900, second October 26, 1998, ship time) are not pressure fluctuations but are caused by inertial motions of the pen as the ship rolled and pitched during this time interval. The largest of these appear on the trace to be

between 0700 (2000, first October 26, 1998, ship time) and 1500 UTC on October 26 (0400, second October 26, 1998, ship time).

79. A detailed segment of the surface weather analysis prepared by Fleet Numerical Weather Central of the U.S. Navy Meteorological and Oceanographic Command at Monterey, California, is shown in Figs. A9 and A10 (not shown here) for 0000 and 1200 UTC on October 26, 1998 (1200, first October 26, 1998, ship time and 0100, second October 26, 1998, ship time).

80. Figure A9 (not shown here) shows the surface weather analysis at 0000 UTC on October 26, 1998 (1200, first October 26, 1998, ship time), with isobars (curved black lines) from 1020 (lower right) to 978 millibars (left, near latitude 41°N, longitude 170°E), wind direction (barbed arrow flies with the wind), and wind speed (here, each full barb on wind arrows equals 10 knots, half barb is 5 knots, and flag is 50 knots).

The position of the M/V *APL China* is shown as a + mark on a plastic overlay for 0000 UTC on October 26, 1998 (1200, first October 26, 1998, ship time).

81. The winds from the U.S. Navy analysis at the position of the vessel (southeast by east at 45–55 knots) match the observed winds in the ship's deck log of southeast by east force 10 = 48–55 knots; the analyzed barometric pressure of 980 millibars is slightly higher than the deck log values of 976 millibars.

82. Figure A10 (not shown here) shows a segment of the U.S. Navy surface weather analysis at 1200 UTC on October 26 (0100, second October 26, 1998, ship time).

83. Between 0000 (1200, first October 26, 1998, ship time) and 1200 UTC (0100, second October 26, 1998, ship time), the low pressure center moved east-northeastward between 25 and 30 knots from near 41°N, 169°E to near 43°N, 173.5°E, while the *APL China* moved slightly south of east from 41.75°N, 170.97°E to 41.43°N, 173.25°E at approximately 23 knots.

84. These combined movements placed the vessel approximately 150 nautical miles southeast of the center of the low pressure system at 1200 UTC on October 26 (0100, second October 26, 1998, ship time), as denoted by the + on the plastic overlay.

85. The analyzed winds on the U.S. Navy weather chart at the vessel's 1200 UTC October 26, 1998 (0100, second October 26, 1998, ship time), position were from the south-southwest at 50 to 55 knots,

and the ship's deck log indicated wind from the south-southwest at force 10 (48–55 knots).

86. The analyzed barometric pressure was 975 millibars, and the ship's logged pressure was 968 millibars at 1200 UTC (0100, second October 26, 1998, ship time) and 988 millibars at 1300 UTC (0200, second October 26, 1998, ship time), bracketing the analyzed 975-millibar value.

87. Waves are generated by the force of wind (air in motion) on the surface of the sea. The height (vertical distance from the lowest point in a trough to the highest point on the adjacent water surface ridge or crests) and the period of waves (time from the passing of one wave's crest to the passing of the next past a fixed point) are determined by the speed of the wind, the duration of its blowing at that speed, and the distance (fetch) over which it has blown at that speed.

88. Table A3 [not shown here; a reproduction of Table 3303 in Bowditch's *American Practical Navigator* (Bowditch 1977)] shows the minimum time T in hours that wind must blow to form waves of H **significant height** (in feet) and P period (in seconds) over a fetch (in nautical miles).

89. "Significant height" is the average height of the highest one-third of the waves present. It is the height most frequently estimated by an observer estimating wave heights.

90. Statistically, the relationship between measured wave heights and those observed are as follows:

| Wave | Relative height |
|---|---|
| Average | 0.6 |
| Significant | 1.00 |
| Highest 10% | 1.29 |
| Highest (one in a thousand) | 1.87 |

91. A 12-meter significant wave height (40 feet) can be generated (Table A3, not shown here) by:
- force 9 wind blowing over a fetch of 700 nautical miles for 49 hours;
- force 10 wind blowing over a fetch of 200 nautical miles for 18 hours; or
- force 11 wind blowing over a fetch of 140 nautical miles for 13 hours.

92. The greatest duration of winds at the location of the M/V *APL China* of force 10 or greater without major change in direction was from 0300 UTC on October 26 (1600, first October 26, 1998, ship time; south by east force 10) to 1500 UTC that day (0400, second October 26, 1998, ship time; south-southwest force 11), an interval of 12 hours.

93. According to Table A3, such a wind would produce wind waves of 37- to 43-foot height with periods of 9 to 10 seconds.

94. Prior to 0300 UTC on October 26 (1600, first October 26, 1998, ship time), wind waves would have come from a southeasterly direction.

95. Subsequent to 1500 UTC on October 26 (0400, second October 26, 1998, ship time), wind waves would have come from westerly and then northwesterly directions.

96. According to Kotsch,[*] "Probably the greatest single item of controversy regarding waves is that of *maximum* wave height, because of the general tendency to underestimate small wave heights and to overestimate large wave heights. Be that as it may, the maximum wave height scientifically recorded thus far (in 1965) in the North Atlantic was 67 feet by a wave recorder aboard an ocean station weather vessel, but this does not mean that higher waves do not occur."

97. According to Kinsman,[†] "If you can keep in mind the fact that really high waves are most uncommon, then it will be safe to entrust you with the following conversational tidbits. These examples of high waves all come from reliable reports. In January, 1894, the *Normania* was damaged . . . by waves at least 40 feet high. The *Ascanius*, during a prolonged gale encountered on the run from Yokahama to Puget Sound, met waves estimated to be at least 70 feet high. The largest wave ever reliably reported had an estimated height of 112 feet. It was encountered on 7 February 1933, during a long stretch of stormy weather, by the U.S.S. *Ramapo* in the Central North Pacific."

98. According to Kinsman,[‡] "It takes a while after a wind springs up to reach the energy balance, a state which may be more briefly termed

---

[*] See Kotsch (1977, p. 219).

[†] See Kinsman (1984, pp. 9–10).

[‡] See Kinsman (1984, p. 8).

a fully aroused sea. If the wind dies out or **changes direction** before the sea is fully aroused, the waves will not develop completely."

99. According to Bowditch,* "The Fleet Numerical Meteorology and Oceanography Center produces synoptic analyses and predictions of ocean wave heights using a spectral numerical model. The wave information consists of heights and directions for different periods and wavelengths. Verification of projected data has proven the model to be very good. Information from the model is provided to the U.S. Navy on a routine basis and is a vital input to the Optimum Track Ship Routing program."

100. Figures A11 and A12 (not shown here) show the significant wave height and primary wave direction as analyzed by the U.S. Navy Fleet Numerical Weather Central for 0000 UTC (1200, first October 26, 1998, ship time) and 1200 UTC on October 26, 1998 (0100, second October 26, 1998, ship time) for the North Pacific Ocean between 35° and 45°N and 170° and 180°E.

101. A clear, plastic overlay has been added to each chart showing the position of the M/V *APL China* at chart time.

102. These analyses indicate the analyzed waves at 0000 UTC October 26, 1998 (1200, first October 26, 1998, ship time), at the position of the M/V *APL China* at that time were 21 feet from the south-southeast.

103. The analysis 12 hours later (at 1200 UTC on October 26, 1998; 0100, second October 26, 1998, ship time) indicates the predominate analyzed waves were 35 feet from the southwest.

104. The U.S. Navy analysis provides information on wave periods as well as wave heights for both "primary" and "secondary" waves at six-hour intervals.

105. The U.S. Navy defines "primary wave period" as the mean wave period for combined wind waves and swells and the "secondary wave period" as the period of the combined "maximum wind waves" and "maximum swell."

106. The "maximum" combined swell and wind waves are defined (in the U.S. Navy model) as the highest one-tenth of the combined waves and swells.

107. The U.S. Navy data for October 26, 1998, were obtained at the analyzed six-hourly times on October 26, 1998, and the values at the successive locations of the *APL China* on that date tabulated.

---

* See Bowditch, ftp://ftp.flaterco.com/xtide/Bowditch.pdf, p. 441.

108. These data are contained in Tab 6 at the end of this report.

109. The values (from the U.S. Navy analysis) for the wave heights and periods at the successive positions of the *APL China* on October 26, 1998, were as follows:

| Time (UTC) | Time (ship time) | Significant wave heights (meters/feet) | Wave periods (seconds) |
|---|---|---|---|
| 0600 | 1900 first October 26, 1998 | 7.3/24 | 10.8 |
| 1200 | 0100 second October 26, 1998 | 11.0/36 | 12.2 |
| 1800 | 0700 second October 26, 1998 | 16.2/53 | 15.3 |
| 2400 | 1300 second October 26, 1998 | 11.5/38 | 14.6 |

110. The U.S. National Oceanographic and Atmospheric Administration (NOAA) provides a map analysis of wave periods over oceanic areas on a three-hourly basis.

111. These are contained in Tab 7 of this report.

112. The NOAA wave heights and periods are a "hindcast" (prepared after the event, rather than operationally concurrent).

113. The significant wave heights and periods from the NOAA hindcast charts in Tab 7 at the successive positions of the *APL China* on October 26, 1998, were as follows:

| Time (UTC) | Time (ship time) | Significant wave heights (meters/feet) | Wave periods (seconds) |
|---|---|---|---|
| 0900 | 2200 first October 26, 1998 | 8.2/29 | 10.7 |
| 1200 | 0100 second October 26, 1998 | 9.3/31 | 11.3 |
| 1500 | 0400 second October 26, 1998 | 11.8/39 | 12.9 |
| 1800 | 0700 second October 26, 1998 | 13.0/43 | 14.0 |
| 2100 | 1000 second October 26, 1998 | 12.9/42 | 14.5 |
| 2400 | 1300 second October 26, 1998 | 11.2/37 | 14.1 |

114. Two polar orbiting satellites [that pass near the north and south terrestrial poles in steeply inclined (relative to the equator) orbits] operationally measure wave heights and estimate ocean surface wind speeds via narrow beam vertically scanning radar.

115. These satellites are the *European Remote Sensing Satellite 2* (ERS-2) and the Ocean Topography Experiment (TOPEX). Their tracks and graphed data for the segment of the North Pacific, including the actual route of the M/V *APL China* and the storm that overtook the vessel on October 26, 1998, are contained in Tab 7—behind the NOAA wave height and period charts—at the end of this report.

116. These satellite data provide a real-time data set useful in calibrating the wind speed and wave height analyses of the U.S. Navy operational wind and wave charts shown in Tab 6 and the NOAA hindcast charts shown in Tab 7.

117. Neither satellite passed directly over the location where the M/V *APL China* was subjected to the highest significant waves on October 26, 1998.

118. The nearest overhead satellite data for the M/V *APL China* high wave encounter were during pass three of satellite *ERS-2* at approximately 2308 UTC on October 26 (1208, second October 26, 1998, ship time), at which time the satellite "observed" wave heights in the range at 11.5 to 14.5 meters (38 to 48 feet).

119. Interpolated values of significant wave heights from the U.S. Navy analysis and the NOAA hindcast at 2300 UTC on October 26, 1998, at the concurrent position of the M/V *APL China* and the *ERS-2* satellite pass three at 2308 UTC (1208, second October 26, 1998, ship time) position were as follows:

| ERS-2 pass three (2308 UTC) | U.S. Navy analysis (2300 UTC) | NOAA hindcast (2300 UTC) |
| --- | --- | --- |
| 11.5–14.5 m | 12.1 m | 11.6 m |

120. By international agreement, ship weather observations are collected and retained in various national archives. The U.S. Archive for Surface Weather Observations from ships at sea is at the U.S. National Climatic Data Center in Asheville, North Carolina.

121. Marine weather observations were archived there for October 1998, including data from the M/V *APL China* and ships of many registries.

122. These are filed by Marsden squares (10° longitude areas over the ocean areas of the world assigned identifying numbers).

123. October 1998 detailed marine weather data within a large area of the western North Pacific examined in detail for this study included those shown in the diagram as follows:

| °N to °N | °W/°E to °W/°E | Marsden square |
|---|---|---|
| 30°–40° | 150°–160°W | 124 |
| 30°–40° | 160°–170°W | 125 |
| 30°–40° | 170°W–180° | 126 |
| 30°–40° | 180°–170°E | 127 |
| 30°–40° | 170°–160°E | 128 |
| 30°–40° | 160°–150°E | 129 |
| 40°–50° | 150°–160°W | 160 |
| 40°–50° | 160°–170°W | 161 |
| 40°–50° | 170°W–180° | 162 |
| 40°–50° | 180°–170°E | 163 |
| 40°–50° | 170°–160°E | 164 |
| 40°–50° | 160°–150°E | 163 |
| 50°–60° | 160°–170°W | 197 |
| 50°–60° | 170°W–180° | 198 |
| 50°–60° | 180°–170°E | 199 |

124. The data from this area for October 25 through 27 are contained in Tab 8 of this report.

125. Examination of those data reveals the highest winds reported from the M/V *APL China* were 50 knots from 130° at 1800 UTC on October 25, 1998 (0600, first October 26, 1998, ship time). These are in agreement with the deck log of the M/V *APL China* for that time.

126. Further examination of the archival marine weather observations from the western North Pacific during October 24–27, 1998, reveals the maximum wind data from any ships were as shown in Table A#.

127. No ship data were found from any ship closer to the center of the rapidly moving low pressure center which affected the M/V *APL China* than the *APL China* at the time of maximum intensity of the low [1200 UTC on October 26 (0100, second October 26, 1998, ship time)]. Table 4 (not included).

128. One ship report (vessel call sign OWEB2 at 30.5°N and 156.2°E) reported a wind from 340° at 97 knots at 1200 UTC on October 25 (2400, October 25, 1998, ship time). This report appears to be in error, as it is not consistent with the weather analyses or other reports.

129. Other than this apparently erroneous report, no ship data were found from any ship with higher winds than the M/V *APL China* (force 11 from 1300 UTC (0200, second October 26, 1998, ship time) to 1600 UTC on October 26, 1998 (0500, second October 26, 1998, ship time).

130. Prior to 0400 UTC on October 25 (1600, October 25, 1998, ship time), all sea state observations on the M/V *APL China* were encoded 5, 6, or 7. These correspond as follows:

| Sea state code | Descriptive term | Approximate height in feet |
|:---:|:---:|:---:|
| 5 | Rough | 8–13 |
| 6 | Very rough | 13–20 |
| 7 | High | 20–30 |

131. From 0400 UTC on October 25 (1600, October 25, 1998, ship time) to 0800 UTC on that date (2000, October 25, 1998, ship time), sea state 8 was logged:

| Sea state code | Descriptive term | Approximate height in feet |
|:---:|:---:|:---:|
| 8 | Very high | 30–45 |

132. Between 1200 UTC on October 25 (2400, October 25, 1998, ship time) and 0600 UTC on October 27 (1900, second October 26, 1998, ship time), all log entries of sea state in the *APL China* deck log were encoded as sea state 9:

| Sea state code | Descriptive term | Approximate height in feet |
|:---:|:---:|:---:|
| 9 | Phenomenal | Over 45 |

133. At 1100 UTC on October 26, 1998 (2400, first October 26, 1998, ship time), the vessel altered course to the southeast as it was "pitching and rolling heavily in confused sea and swell" (per deck log).

134. At that time the winds were changing from southeasterly to southerly to southwesterly as the low pressure storm center moved rapidly northeastward and passed a short distance northwest of the *APL China*.

135. The changing wind directions created newly formed wind waves from the south and southwest atop the previously formed waves (now swells) coming from the southeast and south.

136. These changing wind conditions prevented the development of fully arisen seas from the various directions and resulted in the generation of confused seas with intersecting wave–swell trains creating pyramidal wave forms rather than "orderly" wave trains.

137. Standardized wave statistics are not applicable to the calculation of extreme wave heights in such chaotic sea states.

138. After 0600 UTC on October 27, 1998 (1900, second October 26, 1998, ship time), the sea state encountered by the M/V *APL China* moderated rapidly.

139. The climatic data for representative areas accompanying the isopleth analyses of the U.S. Navy climatic atlas represent the objective compilation of available raw data for the specific area without regard to suspected biases or inconsistencies.

140. Sea level pressure graphics on the climatic charts are not "storm center pressured" but are simple compilations of data from ships that reported within the specified areas.

141. The lowest sea level pressure in the storm of October 26, 1998, at statistical area 23 (closest to the track of the M/V *APL China*), as the storm passed was 1010 millibars.

142. The 1010 millibars in area 23 has an expectancy of 10% in October and 12% in November (see sea level pressure graphics).

143. The text regarding the observations and their processing in the U.S. Navy Marine Climatic Atlas of the World states, "Sea level pressure . . . is one of the least accurate [data sets] because of instrument, coding and conversion errors" (p. 441).

144. Since historically vessels attempted to avoid the more severe storms, there exists a data bias from the exclusion of storm center minimum barometric pressures.

145. The lowest observed sea level pressure associated with the storm of October 24–27, 1998, which affected the M/V *APL China*, were as follows:

964.0 mb at 2300 UTC on October 25 (1100, first October 26, 1998, ship time) at 41.1°N, 168.3°E, with winds from 160° at 46 knots (vessel call sign DGOS); 953.3 mb at 1800 UTC on October 25 (0600, first October 26, 1998, ship time) at 40.2°N, 163.5°E with winds from 080° at 64 knots (vessel call sign WSVR); 968.0 mb at 2200 UTC on October 25 (1000, first October 26, 1998, ship time) at 41.1°N, 168.2°E with winds from 130° at 44 knots (ship call WSVR); and 968 mb from 0900 at 1300 UTC (2200, first October 23, 1998, ship time to 0200, second October 26, 1998, ship time) by the *APL China* near 41.6°N, 176°E with south-southwesterly winds 48 to 63 knots (force 10–11).

146. The presentation by Michael Carr and Lee Chesneau at the Marine Casualties Conference (MARCAS) at the Marine Institute of Technology and Graduate Studies (MITAG) November 14, 2000, stressed that the development, movement, winds, and waves of the storm that overtook the *APL China* on October 26, 1998, were well predicted more than 48 hours in advance.

147. The illustrations and graphics of the Carr–Chesneau presentation are included (together with the Professional Mariner No. 38 article on marine casualties) in Tab 9 at the back of this report.

148. In his presentation of "Interpretation and Application of Meteorological Information for the Licensed Mariner," the Keystone Shipping Company (in 1997) Captain Austin Dooley outlined procedures for mariners to be aware of severe weather prospects.

149. Captain Dooley's presentation reprints are included in Tab 10 at the back of this report.

150. The development and movement of the storm of late October 1998 that impacted the *APL China* was characteristic of the procedures outlined in Captain Dooley's presentation.

D. **OPINIONS:**

1. Before leaving Taiwan, the weather risks for the planned voyage to Seattle could be best assessed by reference to climatological data, such as contained in the *U.S. Navy Climatic Atlas of the Oceans, Volume 2: North Pacific.*

2. Climatological data for October indicate a voyage along the track of the M/V *APL China* will pass through a "principal area of cyclogenesis" over the western North Pacific early in the voyage.

3. Rapidly developing low pressure centers (storms) in that area have a climatological expectancy of moving northeastward at speeds of 30–35 knots.

4. The area traversed by the M/V *APL China* west of longitude 180° is one in which the development of explosively intensifying lows are most likely to occur.

5. The storm of October 24–27, 1998, which overtook and passed the M/V *APL China* on October 26, was a minimal "Bergeron bomb."

6. There is an expectancy of four "bombs" in 10 days in the Pacific in late October, most of which occur along the preferred storm track in Fig. A6 (not shown here), crossed by the actual route of the M/V *APL China*.

7. During the course of the voyage, the climatological assessment of weather risks could be updated for 48 to 96 hours in advance by weather analyses and forecasts issued by the government or private organizations such as the JMA, the U.S. NWS, and private organizations.

8. The development and movement of the October 24–27, 1998, storm affecting the M/V *APL China* was well identified and well forecast by the Japan Meteorological Agency and the U.S. National Weather Service, both of whose analyses and forecasts were broadcast for use by ships in the western North Pacific Ocean.

9. As the storm overtook and passed the M/V *APL China* on October 26, 1998, (UTC date), the vessel encountered the maximum winds and highest waves within the storm circulation, southeast of the northeastward-moving storm center.

10. Winds speeds reached Beaufort force 11 (56–63 knots) in the maximum storm winds from the south, becoming southwest and west, affecting the M/V *APL China* between 1300 and 1600 UTC on October 26, 1998 (0200 and 0500, second October 26, 1998, ship time).

11. There was no force 12 (64 knots or greater) associated with this storm.

12. The probable wave heights and periods encountered by the *APL China* on October 26, 1998, increased from 29 feet and 10.7 seconds at 0900 UTC (2200, first October 26, 1998, ship time) to 43 feet and 14 seconds at 1900 UTC (0700, second October 26, 1998, ship time) and then decreased as follows:

**Probable wave conditions encountered by the M/V *APL China***

| Time (UTC) | Time (ship time) | Significant wave heights (feet) | Wave periods (seconds) |
|---|---|---|---|
| 0900 | 2200 first October 26, 1998 | 29 | 10.7 |
| 1000 | 2300 first October 26, 1998 | 29 | 10.9 |
| 1100 | 2400 first October 26, 1998 | 30 | 11.1 |
| 1200 | 0100 second October 26, 1998 | 31 | 11.3 |
| 1300 | 0200 second October 26, 1998 | 33 | 11.8 |
| 1400 | 0300 second October 26, 1998 | 36 | 12.4 |
| 1500 | 0400 second October 26, 1998 | 39 | 12.4 |
| 1600 | 0500 second October 26, 1998 | 41 | 13.3 |
| 1700 | 0600 second October 26, 1998 | 42 | 13.6 |
| 1800 | 0700 second October 26, 1998 | 43 | 14.0 |
| 1900 | 0800 second October 26, 1998 | 43 | 14.2 |
| 2000 | 0900 second October 26, 1998 | 42 | 14.4 |
| 2100 | 1000 second October 26, 1998 | 42 | 14.5 |
| 2200 | 1100 second October 26, 1998 | 40 | 14.4 |
| 2300 | 1200 second October 26, 1998 | 38 | 14.2 |
| 2400 | 1300 second October 26, 1998 | 37 | 14.1 |

13.  The sea conditions in the southeast quadrant of the storm (those encountered by the M/V *APL China* on October 26, 1998) were "confused" or "chaotic," with interference between waves from the various directions (southeast, south, southwest, and west).

14.  The changing wind directions (from southeast to south and southwest to west) prevented the development of a fully risen sea from any of the changing wind directions.

15. Under those conditions, the relationship between average, significant, highest 10%, and highest waves are less applicable than in the
conditions of a persistent, steady wind blowing from a constant
direction over a long fetch for many hours.

E. **BASIS FOR OPINIONS:**

These opinions are based on the facts in section C (above), the data
and information relied upon in section F (below), and the education
training, knowledge, and experience of the witness.

F. **DATA AND INFORMATION RELIED UPON:**

1. Northern Hemisphere surface charts for the North Pacific
   from 120°E to 120°W and 10°–80°N for 1200 UTC on October
   23 through October 28, 1998, from the National Climatic Data
   Center.
2. Ship reports from the North Pacific for 160°E to 160°W and
   30°–60°N for 1200 UTC on October 23, 1998, through October 28,
   1998, from the National Climatic Data Center.
3. U.S. Navy surface analysis and wave data analysis for October 26,
   1998.
4. Satellite data on North Pacific wind and waves for October 25–27,
   1998, from TOPEX and *ERS-2* polar orbiting satellites.
5. NOAA "Wave Watch III" maps of wave heights and periods for
   October 26, 1998.
6. High seas forecasts pertinent to the North Pacific covering 160°E
   to 160°W and 30°–50°N from the National Climatic Data Center.
7. Report on the weather, waves, and climatology concerning the 26
   October, 1998, M/V *APL China* prepared by Dooley Seaweather
   Analysis, Inc., and Oceanweather, Inc.
8. Report No. 15575-2-SMB: Seakeeping tests for the *APL China*, final
   data report, September 2000 prepared by Netherlands Maritime
   Research Institute.
9. Report No. 15575-3-SMB: Additional seakeeping tests for the *APL
   CHINA*, data report, September 2000 by Netherlands Maritime
   Research Institute.
10. Report by Herbert Engineering Corp.: "M/V *APL China*—Voyage
    30E causes of container loss and damage" (September 2000).
11. APL 015616-15634 Polaris printouts made during the heavy
    weather incident.

12. M/V *APL China* deck log book, October 21, 1998, through October 31, 1998.
13. M/V *APL China* data log, starting October 26, 1998.
14. M/V *APL China* data bell book, starting October 26, 1998.
15. National Weather Service barogram from M/V *APL China* for 20–28 October.
16. Weathernews records directly relevant to the voyage in issue: voyage report and weather charts for October 22–27, 1998.
17. WNI records relevant to *APL China* voyage 30E of October–November 1998, including the following:
    * voyage report,
    * route chart,
    * vessel performance profile,
    * message traffic, and
    * weather charts for October 22–27, 1998.
18. Defendant deposition exhibits as follows:
    * 158: NWI message to master of *APL China* of October 20, 1998, regarding route, weather forecast, typhoon prediction, and expected conditions.
    * 242: NWI message to master of *APL China* regarding routing and forecasts October 22–29, 1998.
    * 243: National Weather Service synopses, forecasts, and warnings; Hong Kong Meteorological Service warnings; and Japan Meteorological Agency warnings on Typhoon BABS for various dates between October 13 and 30, 1998.
19. Defendants deposition exhibits 245: weather faxes from October 13 to 30, 1998, showing surface weather analysis for the North Pacific made by the Japan Meteorological Agency and weather reports via Inmarsat from October 12 to 30, 1998, showing North Pacific surface weather maps analyses generated by the U.S. National Weather Service.
20. Data and information regarding prognostic weather charts for dates in October 1998 relative to the track of *APL China* generated by WNI-Oceanroutes, voyage reports, and message traffic relative to the *APL China* in September and November 1998; North Pacific High Seas forecasts prepared by the U.S. National Weather Service for October 23 through 27, 1998; surface pressure Northern Hemisphere map analyses and prognostics at 23- to 24-hour intervals 0 through 144 hours for October 24 through 29, 1998; and segments

of surface weather analysis charts for the North Pacific, generated by WNI-Oceanroutes October 25 through 27, 1998.

21. Part of deposition of Parvez P. Guard.

22. Deposition of Staw Guan Lee.

23. Deposition of Mohamed Badruddin Bin Mahamed Yusoff.

24. American Bureau of Shipping Research and Development Division, Technical Report RD-87027: "Acceleration and Roll Motion Analysis for APL C10 Class Containerships," prepared by Herbert Engineering Corporation, November 1987.

25. Houghton, D. D., Ed., 1985: *Handbook of Applied Meteorology.* Wiley, 1461 pp.

26. Kotsch, W. J., 1977: *Weather for the Mariner.* 2nd ed. Naval Institute Press, 272 pp.

27. Donn, W. L., 1975: *Meteorology.* 4th ed. McGraw-Hill, 608 pp.

28. *International Symposium on Ocean Wave Measurement and Analysis,* Vol. 1, New Orleans, LA, American Society of Civil Engineers.

29. Cotton, W. R., and R. A. Anthes, 1989: *Storm and Cloud Dynamics.* Academic Press, 883 pp.

30. Hartmann, D. L., 1994: *Global Physical Climatology.* Academic Press, 411 pp.

31. Sanders, F., and J. R. Gyakum, 1980: Synoptic-dynamic climatology of the "bomb." *Mon. Wea. Rev.,* **108**, 1589–1606, doi:10.1175/1520-0493(1980)108 <1589:SDCOT>2.0.CO;2.

32. Huschke, R. E., Ed., 1959: *Glossary of Meteorology.* 1st ed. Amer. Meteor. Soc., 638 pp.

33. Kinsman, B., 1984: *Wind Waves: Their Generation and Propagation on the Ocean Surface.* Dover, 676 pp.

34. Bowditch, N., 1977: *American Practical Navigator.* Vol. 1. Defense Mapping Agency Hydrographic Center, 1386 pp.

35. U.S. Navy, 1977: *U.S. Navy Marine Climatic Atlas of the World, Volume II: North Pacific Ocean.* Naval Oceanography and Meteorology.

36. "Interpretation and Application of Meteorological Information for the Licensed Mariner," presented by Keystone Shipping Company, in May 1997 by Dooley Seaweather Analysis, Inc.

37. Presentation at *Professional Mariner's Maritime Casualties Conference and Expo MarCas 2000* program: "APL CHINA" by Michael Carr [Marine Institute of Technology and Graduate Studies (MITAGS) faculty] and Lee Chesneau (National Weather Service Marine Prediction Center) November 14, 2000.

38. Professional Mariner, 1999: "Maritime Casualties: Pacific Weather 'bomb' results in massive container loss." *Prof. Mariner*, **38**, 63–64.

## G. EXHIBITS:

It is anticipated that maps, charts, and graphs portraying some of the information in the facts in section C above, the data and information in section F above, and stated in the opinions in section D above will be utilized to clarify and demonstrate the testimony of this witness.

## H. RECENT PUBLICATIONS BY THE WITNESS INCLUDE:

Papers authored by the witness since 1980 include the following:

Bohan, W. A., W. H. Haggard, and W. L. Maynard, 1987: Visual representation of combined weather and aircraft data for litigation. *Proc. Third Int. Conf. on Interactive Information and Processing Systems for Meteorology, Oceanography and Hydrology*, New Orleans, LA, Amer. Meteor. Soc., 188–189.

Falconer, P. D., and W. H. Haggard, 1990: Forensic meteorology. Forensic Sciences, Matthew Bender and Company, Chapter 35B 1–48

Falconer, P. D., and W. H. Haggard, 1999: Forensic meteorology. *Forensic Sciences*, Matthew Bender and Company, 35B-1–45.

Haggard, W. H., 1980a: Radar imagery for post analysis of mountain valley flash floods. Preprints, *Second Conf. on Flash Floods*, Atlanta, GA, Amer. Meteor. Soc., 97–100.

Haggard, W. H., 1980b: What's in a wind study? *Proc. Second Joint Conf. on Industrial Meteorology*, New Orleans, LA, Amer. Meteor. Soc., 825–831.

Haggard, W. H., 1980c: Some micro-economic aspects of applied climatology. *Proc. Conf. on Climatic Impacts and Societal Response*, Milwaukee, WI, Amer. Meteor. Soc.

Haggard, W. H., 1981: Applied climatology: Some data sources and applications. *J. Appl. Meteor.*, **20**, 1412–1414.

Haggard, W. H., 1983: Weather after the event. *Proc. Ninth Conf. on Aerospace and Aeronautical Meteorology*, Omaha, NE, Amer. Meteor. Soc., 139–142.

Haggard, W. H., 1984: The AMS certified consulting meteorologist program. *Proc. 14th Conf. on Broadcast Meteorology*, Clearwater Beach, FL, Amer. Meteor. Soc.

Haggard, W. H., 1985: Meteorologists as expert witnesses. *Proc. 15th Conf. on Broadcast Meteorology*, Honolulu, HI, Amer. Meteor. Soc.

Haggard, W. H., and S. M. McCown, 1985: Architectural and structural design. *Handbook of Applied Meteorology*, D. D. Houghton, Ed., John Wiley and Sons.

Haggard, W. H., 1987: Climatic information for boaters. *The Ensign*, U.S. Power Squadrons, 21–22.

Haggard, W. H., 1989a: Weather testimony in litigation. *Proc. Third Int. Conf. on the Aviation Weather System*, Anaheim, CA, Amer. Meteor. Soc., 448–450.

Haggard, W. H., 1989b: Urban weather. Int. *J. Environ. Stud.*, **36**, 73–82.

Haggard, W. H., 1991: Climate services. *Climate in Human Perspective: A Tribute to Helmut E. Landsberg*, Kluwer Academic, 85–90.

Haggard, W. H., 1992a: Boating weather in USPS District 17. *Asheville NC Power Squadron*, U.S. Power Squadrons, 26 pp.

Haggard, W. H., 1992b: Private sector use of climate information. *Proceedings of 1992 Southeast Climate Symposium*, Southeast Regional Climate Center.

Haggard, W. H., 1994: Conquer nature's challenges through weather consciousness. *The Ensign*, U.S. Power Squadrons.

Haggard, W. H., and S. W. Smutz, 1994: Forensic meteorology. *Proc. Seventh Aviation Law/Insurance Symp.*, Daytona Beach, FL, Embry

Riddle Aeronautical University, 1–12.

Haggard, W. H., 1996a: Listen to the data they have a story to tell. *Proc. Symp. on Environmental Applications*, Atlanta, GA, Amer. Meteor. Soc., 15–21.

Haggard, W. H., 1996b: Thunderstorms, microbursts, aircraft accidents and litigation. *National Weather Association National Meeting*, Cocoa Beach, FL, National Weather Association.

Haggard, W. H., 2000a: Millennium Perspectives on Better Prediction of High-Impact Weather Events—"Prediction of High Impact Weather Events." *Bull. Amer. Meteor. Soc.*, **81**, 102.

Haggard, W. H., 2000c: Are the data you rely on valid? *Proc. 12th Conf. on Applied Climatology*, Asheville, NC, Amer. Meteor. Soc., 3a.4. [Available online at https://ams.confex.com/ams/May2000/techprogram/paper_13570.htm.]

Haggard, W. H., 2003: The demands of forensic climatology. *Handbook of Weather, Climate and Water: Dynamics, Climate, Physical Meteorology, Weather Systems, and Measurements*, McGraw Hill, 829–839.

I.  **COMPENSATION:** Costs have been, are, and will be billed at the prevailing published fee schedule for Climatological Consulting Corporation.

J.  **PAST EXPERT TESTIMONY:**
    Other cases in which the witness has testified at trial or deposition in recent years include the following:

| Date of Event | Location | Attorney | Depo/Trial Date | Style | Court |
| --- | --- | --- | --- | --- | --- |

*[The next 10 pages of the expert witness statement listed information on cases from 1994 to 2000 from St. Croix in the Virgin Islands to San Diego, California, and Boston, Massachusetts, with 35 depositions and 15 in-court trial testimony. These tabulations provided the information needed by the involved lawyers to request transcripts of all of the expert's testimony in those seven years of prior trials.]*

K. **<u>SUPPLEMENTATION OF OPINIONS:</u>** The opinions I have stated are subject to supplementation depending upon the availability of additional information, such as further deposition testimony, reports of other expert witnesses, and additional documentation which may become later available.

_____

William H. Haggard, C.C.M.

# ABOUT THE AUTHOR

Fascinated by the weather from early childhood, the author had a long career as a meteorologist with the U.S. Navy in WWII and the Korean conflict, retiring as a captain. He served as a forecaster and researcher in the U.S. Weather Bureau. He was a college instructor at North Carolina State University and a climatologist before serving as assistant chief in the Office of Plans for the U.S. Weather Bureau and then as deputy director and director to the now National Climatic Data Center. After retiring in 1975, he formed Climatological Consulting Corporation, which provided services in forensic meteorology.

Bill graduated from Yale in 1942 with a degree in Physics, the Massachusetts Institute of Technology with a Certificate of Professional Meteorology, the University of Chicago with an M.S. in Meteorology, and had two years of further graduate training at the Florida State University.

During his careers, the author published technical articles in professional journals on long-range forecasting, hurricane and tropical storm prediction, excessive rainfall probabilities, applications of climatic data to operational programs, climatic analyses, flash floods, wind studies, forensic meteorology, radar and satellite meteorology, and so on.

The author has served as an associate editor of *Journal of Applied Meteorology*, consulting editor of *Weatherwise*, and editorial board member of the American Meteorological Society *Glossary of Meteorology* (second edition).

The author has also served on international assignments as chairman or member of several U.S. delegations and working groups of the World Meteorological Organization and of the Intergovernmental Oceanographic Commission.

Bill is a Certified Consulting Meteorologist and has been elected as a fellow and honorary member of the American Meteorological Society.

He has been affiliated with the American Institute of Aeronautics and Astronautics, National Council of Industrial Meteorologists, American Association for Advancement of Science, American Geophysics Union, American Meteorological Society, International Oceanographic Foundation, New York Academy of Science, United States Power Squadron, American Association of State Climatologists, American Association of Weather Observers, and National Weather Association.

# REFERENCES

AMS, 2012a: Cumulonimbus. Glossary of Meteorology, http://glossary.ametsoc.org/wiki/Cumulonimbus.

AMS, 2012b: Whiteout. Glossary of Meteorology, http://glossary.ametsoc.org/wiki/Whiteout.

AMS, 2012c: Temperature inversion. Glossary of Meteorology, http://glossary.ametsoc.org/wiki/Temperature_inversion.

AMS, 2012d: Sea state. Glossary of Meteorology, http://glossary.ametsoc.org/wiki/Sea_state.

AMS, 2012e: Comfort zone. Glossary of Meteorology, http://glossary.ametsoc.org/wiki/Comfort_zone.

Belew, D. O., 1989a: Memorandum opinion. Air Crash at Dallas/Fort Worth Airport on August 2, 1985. Kathleen CONNORS, et al. v. United States of America, MDL Docket 657 CA4-87-060-K, U.S. District Court for the Northern District of Texas Fort Worth Division 1989, 72 pp., http://www.leagle.com/decision/19891978720F Supp1258_11793/IN%20RE%20AIR%20CRASH%20AT%20DALLAS/FORT%20WORTH%20AIRPORT.

Belew, D. O., 1989b: Judgment. Air Crash at Dallas/Fort Worth Airport on August 2, 1985. Kathleen CONNORS, et al. v. United States of America, MDL Docket 657 Civil Action 4-87-060-K (Consolidated with Civil Action 4-87-139-K), U.S. District Court for the Northern District of Texas Fort Worth Division 1989, 1 p.

Bureau of Indian Affairs, U.S. Department of the Interior, Indian Affairs, History of BIA, http://www.bia.gov/WhoWeAre/BIA/.

Bowditch, N., 1977: *American Practical Navigator*. Vol. 1. Defense Mapping Agency Hydrographic Center, 1386 pp.

Bradley, M. D., 1983: *The Scientist and Engineer in Court*. American Geophysical Union, 111 pp.

Cantor, B. J., 1996: *The Role of the Expert Witness in a Court Trial (A Guide for the Expert Witness)*, 160 pp.

Carr, G. G., 1982a: United States District Court Middle District of Florida Tampa Division, In the Matter of the Complaint of HERCULES CARRIERS, INC., for exoneration from or limitation of liability as owners of the M/V SUMMIT VENTURE, Case No. 80-563-Civ-T-GC, 5 pp.

Carr, G. C., 1982b: Partial Summary Judgment, Appendix C, 566F.Supp.962, pp 30.

Cole, L. A., 1988: *Clouds of Secrecy: The Army's Germ Warfare Tests over Populated Areas*. Rowman and Littlefield, 188 pp.

Cummings, J., Fetherston, D., 1976: Army: *Germ warfare was 'tested' on U.S. cities*. San Francisco Chronicle, December, 22, 1976.

Curtis, J., 1980: *Skyway May 9, 1980: The Day the Sky Fell*. Chillum, 52 pp.

FAA, 1983: Federal Aviation Administration, Advisory Circular, Subject: Thunderstorms, AC No: 00-24B, 7 pp., http://www.faa.gov/documentLibrary/media/Advisory_Circular/AC_00-24B.pdf.

Federal Aviation Administration, 2015: FAA aviation weather cameras. Accessed July 29, 2016, http://avcams.faa.gov.

Fujita, T. T., 1985: *The Downburst: Microburst and Macroburst*. Satellite and Mesometeorlogy Research Project, Department of the Geophysical Sciences, University of Chicago, 122 pp.

Fujita, T. T., 1986: *DFW Microburst on August 2, 1985*. University of Chicago, 154 pp.

Garrison, P., 2009: Dark passage: A night VFR rescue mission deteriorates into night IMC. *Flying Magazine*, 8 August, http://www.flyingmag.com/safety/accident-investigations/dark-passage.

Ginsberg, S., 1998: Lawsuits rock APL's boat. *San Francisco Business Times*, 22 November 1998, http://www.bizjournals.com/sanfrancisco/stories/1998/11/23/story1.html.

Glickman, T., Ed., 2000: *Glossary of Meteorology*. 2nd ed. Amer. Meteor. Soc., 855 pp., http://glossary.ametsoc.org/.

Grace, E. V., 1981: Findings and facts. Summit Venture collision with Sunshine Skyway, U.S. Coast Guard, Department of Transportation Rep., 27 pp., http://www.uscg.mil/hq/cg5/cg545/docs/boards/summitventure.pdf.

Greene, T., 1994: Webb v. United States. 840 F. Supp. 1484, Civ. No. 90-C-625G, 90-C-826J (D. Utah January 10, 1994), 70 pp.

Haggard, W. H., Harrison, H. T., Myers, V. A., and Baum, W. A., 1986: A Report on Climatic Factors, History and Variability Affecting the White Mountain Apache Indian Reservation, Arizona, by Climatological Consulting Corporation, Defense Exhibit F-1, 123 pp.

Haggard, W. H., McVehil, G. E., 1993: A Report to Counsel on the Causes and Origin of Fog at Exit 36 on Interstate Highway 75 on December 11, 1990, April, 1993, 167 pp.

Haggard, W. H., Smutz, S. W., 1996: A Preliminary Report on The Weather of April 23, 1991 Affecting the Georgia Pacific Mill at Palatka, Florida, January 1996, 21 pp.

Huschke, R. E., Ed., 1959: *Glossary of Meteorology*. 1st ed. Amer. Meteor. Soc., 638 pp.

Johnson, K. G., F. Quinones-Marquez, and R. Gonzalez, 1987: Hydraulic analyses of water-surface profiles in the vicinity of the Coamo Dam and Highway 52 Bridge, southern Puerto Rico: Flood analyses as related to the flood of October 7, 1985. U.S. Geological Survey, Water-Resources Investigations Rep. 87-4039, 32 pp., https://pubs.er.usgs.gov/publication/wri874039.

Kareem, A., Ed., 1985: *Hurricane Alicia: One Year Later*. American Society of Civil Engineers, 335 pp.

Kinsman, B., 1984: *Wind Waves: Their Generation and Propagation on the Ocean Surface*. Dover, 676 pp.

Kotsch, W. J., 1977: *Weather for the Mariner*. 2nd ed. Naval Institute Press, 272 pp.

Meisner, B., 1989: Deposition of Bernard Meisner, Ph.D., No. 85CV1244, Texas City Refining, Inc. vs GT&MC, Inc., F/K/A Graver Tank & MFG., Co., Inc., A Subsidiary of Aerojet-General Corporation, in the District Court of Galveston County, Texas, 122nd Judicial District, 59 pp.

National Hurricane Center, 2015: What is UTC or GMT time? NOAA National Hurricane Center, http://www.nhc.noaa.gov/aboututc.shtml.

NCDC, 1991: *Storm Data*. Vol. 33 (April 1991), 240 pp., https://www.ncdc.noaa.gov/IPS/sd/sd.html.

Nettesheim, C. C., 1987: Opinion, In the United States Claims Court, No. 22-H, (filed February 6, 1987), White Mountain Apache Tribe of Arizona, Plaintiff v. The United States, Defendant, Indian Claims; Indian Claims Commission, Act; claims for mismanagement of water, rangeland, and timber resources; continuing wrong; damages, 103 pp.

NTSB, 1973: Aircraft accident report: Pan Alaska Airways, Ldt., Cessna 310C, N1812H: Missing between Anchorage and Juneau, Alaska, October 16, 1972. National Transportation Safety Board Aircraft Accident Rep. NTSB-AAR-73-01, 25 pp., http://www.ntsb.gov/investigations/AccidentReports/Reports/AAR7301.pdf.

NTSB, 1981: Marine accident report: Ramming of the Sunshine Skyway Bridge by the Liberian bulk carrier Summit Venture, Tampa Bay, Florida, May 9, 1980. National Transportation Safety Board Marine Accident Rep. NTSB/MAR-81/03, 55 pp.

NTSB, 1986: Aircraft accident report: Delta Air Lines, Inc., Lockheed L-1011-385-1, N726DA, Dallas/Fort Worth International Airport, Texas, August 2, 1985. National Transportation Safety Board Aircraft Accident Rep. NTSB/AAR-86/05, 167 pp., http://www.ntsb.gov/investigations/AccidentReports/Reports/AAR8605.pdf.

NTSB, 1992: Highway accident report: Multiple-vehicle collisions and fire during limited visibility (fog) on Interstate 75 near Calhoun, Tennessee, December 11,

1990. National Transportation Safety Board Highway Accident Rep. NTSB/HAR-92/02, 84 pp., http://www.ntsb.gov/investigations/AccidentReports/Reports/HAR9202.pdf.

NTSB, 1995: Factual Report Aviation, National Transportation Safety Board, Rep. NTSB ID: MIA95FA145, 5 pp.

NWS, 1991: Special Weather Statements issued by Daytona Beach and Jacksonville, Florida during the period of 5 a.m. through 2 p.m. on April 23, 1991.

Paul, M. M., 1998: Finding of Fact and Conclusions of Law in the United States District Court for the Northern District of Florida Gainesville Division, Case No. 1:97CV111-MMP, pp.27.

PBS Online, 2001: Wonders of the World Databank. http://www.pbs.org/wgbh/buildingbig/wonder/structure/sunshine_skyway.html.

Poynter, D., 1987: *The Expert Witness Handbook: Tips and Techniques for the Litigation Consultant*. Para Publishing, 243 pp.

Salottolo, G., D., 1995: Meteorological Factual Report, National Transportation Safety Board, Office of Aviation Safety, September 27, 1995, 4 pp.

Sanders, F., and J. R. Gyakum, 1980: Synoptic-dynamic climatology of the "bomb." *Mon. Wea. Rev.*, **108**, 1589–1606, doi:10.1175/1520-0493(1980)108<1589:SDCOT>2.0.CO;2.

Tannehill, I. R., 1944: *Hurricanes: Their Nature and History, Particularly Those of the West Indies and the Southern Coasts of the United States*. Princeton University Press, 269 pp.

U.S. Chemical Corps Biological Laboratories, 1951: Special Report No. 142: *Biological Warfare Trials at San Francisco, California, 20-27 September 1950*.

U.S. Office of Climatology, 1959: *North Pacific Ocean*. Vol. 2, *Climatological and Oceanographic Atlas for Mariners*, U. S. Government Printing Office.

Weiss, L.D., 2004: Collision on I-75. American Public Health Association, 109 pp.

Wentworth, R. J., 1995: Air Traffic Control Group Chairman's Factual Report of Investigation, NTSB No: MIA-95-FA-145, 17 pp.

Wheat, R. P., A. Zuckerman, and L. A. Rantz, 1951: Infection due to chromobacteria. *AMA Arch. Intern. Med.*, **88**, 461–466, doi:10.1001/archinte.1951.03810100045004.

White Mountain Apache Land Restoration Code, 4/08/98. http://www.wmat.nsn.us/Legal/Land%20Restoration%20Code.pdf

# INDEX

fog
  in auto accidents (*see Downing v. Bowater*)
  in aviation, 2, 86–87, 91, 131–133
  definition and types of, 100–101, 105
  detection and warning system, 116
  in marine navigation (*see* Sunshine Skyway Bridge collapse)
forensic meteorology, 153–154
Forest Management Plans, 77
Fort Apache Indian Reservation, 72
Fowler, White, Gillen, Boggs, Villareal, and Banker, 26
Franklin, Benjamin, 73
frontal fog, 100
Fujita, Tetsuya "Ted" (Dr.), 37–38, 47, 48–49, 54–55, 96, 154
Fujita tornado scale, 93
funnel clouds, 95

Gainesville, FL. *See McNair v. USA*
Galveston, TX. *See* Hurricane Alicia
*Garrell/Bike v. State of Connecticut*, 95–96
Georgia-Pacific paper mill case, 117–123
  background, 117–118
  conclusions, 122–123
  site visit and data gathering, 118–122
  weather surveillance radar, 120–122
germ warfare. *See Nevin v. USA*
Gibbs, Sanford, 6
Gibson, Robert L., 133
Gonzalez, Alex (Atty.), 65
Gonzalez, Jose (Hon.), 56, 57
*Goodsailor* (motor vessel), 23
Graf, William (Dr.), 79
Green, John (Maj.), 72
Greene, Thomas (Hon.), 91–92
Greyhound Bus Lines, 26, 32
ground fog, 100

Guillermety, Ortiz and Associates Engineering Co. (GTA), 64
Gulf States Weather, 29

Hagans, Charles, viii, 4, 6–9
Haggard, William L., career overview, 153–155
hail, 94
Hanlon, David, 27
Hanna, Steven (Dr.), 108
Hardin, James (Atty.), 147–151
Harned, Stephen, 147–148
Harrison, Henry, 27, 78, 79
Haught, Charles (Hon.), 142
Hayden, Raymond (Atty.), 142, 144–145
Hayes, Andy, 89
Healy and Baillie, 142
Hercules Carriers, Ltd., 26, 32
HF cyclone (hurricane force extratropical cyclone), 137
Hill, Benjamin, 27
Hill, Rivkins, and Hayden, 142
Hiwassee River (TN), 101, 102, 105, 106, 109–110
Hodges, Doughty, and Carson, 104
Howard, Needles, Tammen, and Bergendoff (HNTB), 65
Hsiung Chu Liu (Capt.), 22–23
Hudson, Henry Orlando (Hon.), 150
Hurricane Alicia, 35–40, 154
  event details, 35–37
  litigation, 37–40
  trial and judgments, 39–40
*Hurricane Alicia: One Year Later* (Kareem), 36
hurricanes
  Alicia (*see* Hurricane Alicia)
  Puerto Rico, 62
  Saffir–Simpson hurricane scale, 36
  San Felipe II, 69